Design, Modeling and Characterization of Bio-Nanorobotic Systems

Mustapha Hamdi • Antoine Ferreira

Design, Modeling and Characterization of Bio-Nanorobotic Systems

Springer

Dr. Mustapha Hamdi
Institut PRISME
ENSI Bourges
88 Boulevard Lahitolle
18020 Bourges
France
mfhamdi@gmail.com

Dr. Antoine Ferreira
Institut PRISME
ENSI Bourges
88 Boulevard Lahitolle
18020 Bourges
France
antoine.ferreira@ensi-bourges.fr

ISBN 978-90-481-3179-2 e-ISBN 978-90-481-3180-8
DOI 10.1007/978-90-481-3180-8
Springer Dordrecht Heidelberg London New York

Library of Congress Control Number: 2010937677

Cover design: eStudio Calamar S.L.

Printed on acid-free paper

Springer is part of Springer Science+Business Media (www.springer.com)

To my family, on whose constant encouragement and love I have relied throughout my time at the Academy. I am grateful also to the examples of my brother, It is to them that I dedicate this work.

Acknowledgements

I would like to thank Prof. Y. Touré for giving me opportunities to make the most out of my thesis years in the PRISME institute. I would like to thank Prof. A. Ferreira my thesis advisor for his scientific guiding and suggestions through the thesis work.

It is a pleasure to thank the reviewers of my thesis: Prof. B. Courtois and Prof. S. Fatikow for their detailed reports on this thesis. The quality of the manuscript is definitively improved thanks to their comments. My sincere gratitude also goes to the rest of my thesis committee: Prof. B. Nelson and Prof. A. Voda.

A special thank and appreciation to Professor Constantinos Mavroidis, Professor Director of the Biomedical Mechatronics Laboratory Department of Mechanical and Industrial Engineering, for his scientific guiding and suggestions during the fruitful collaboration about Bio-nanodevices prototyping and characterization.

It is difficult to overstate my appreciation to Prof. Bradley Nelson, supervisor of the Multi-Scale Robotics Laboratory in the ETH of Zurich with whom I began to learn about experimentation and fabrication of carbon nanotubes based nanodevices. Not only a great mentor, he has also been a cornerstone in my professional development. I have also to thank the Multi-Scale Robotics Laboratory team: Dr. Lixin dong, Dr. Arun Subramanian and Ph.D. student Kaiyu Shou.

People in my institute helped me during my scientific explorations. I feel privileged to thank all of them and particularly the following researchers for both their insightful comments and the enjoyable discussions: A. Ben Ali and V. Idasiak. Of course, this research would not have been possible without the financial support of the FSE and the Region of Cher. For these I would like to acknowledge the financial support from the ENSI of Bourges.

Contents

List of Figures

List of Tables

Introduction

Nanorobotics is an emerging area of scientific and technological opportunity. It is a new and rapidly growing interdisciplinary field addressing the assembly, construction and utilization of molecular devices based on nanoscale principles and/or dimensions. The size-related challenge is the ability to measure, manipulate, and assemble matter with features on the scale of 1–100 nm. In order to achieve cost-effectiveness in nanotechnology it will be necessary to automate molecular manufacturing. The engineering of molecular products needs to be carried out by robotic devices, which have been termed nanorobots. A nanorobot is essentially a controllable machine at the nanometer or molecular scale that is composed of nano-scale components and algorithmically responds to input forces and information. Bio-nanorobotics, namely biomolecular robots, represents a specific class of nanorobots where proteins and DNA could act as motors, mechanical joints, transmission elements, or sensors. If all of these different components were assembled together they can form bio-nanorobots with multi-degree-of-freedom, able to apply forces and manipulate objects in the nanoscale world. These bio-components seem to be a very logical choice for designing nanorobots. In addition since some of the core applications of nanorobots are in the medical field, using bio-components for these applications seems to be a good choice as they both offer efficiency and variety of functionality. This idea is clearly inspired by nature's construction of complex organisms such as, bacteria and viruses which are capable of movement, sensing and organized control.

The design of nanorobotic systems requires the use of information from a vast variety of sciences ranging from quantum molecular dynamics to kinematic analysis. So far, there does not exist any particular guideline or a prescribed manner that details the methodology of designing a bio-nanorobot. As research development is at the interface of physical sciences and biology it requires multi-skilled teams. To achieve this it is essential for the future molecular roboticians to be able to design and prototype the bio-nanomechanics, to develop dynamic and kinematic models, to study their dynamic performances and to optimize their structure. Although many of the described technologies have been developed into more or less mature products for robots acting in the macro-world, the nano-size of the molecular robots poses extreme challenges and requires a complete rethinking of the design and prototyping methodologies.

Based on the general research issues of design, modeling and prototyping, we have proposed the following research agenda for this thesis:

Chapter 1 presents a state-of-the-art in the emerging field of nano and bio-nano-robotics and its applications. First, it discusses some of the essential properties and dynamical laws which make this field more challenging and unique than its macro-scale counterpart. Second, we review the relation between molecular behavior and system/device response, by analyzing the current trend in modeling and simulation tools, which can analyze phenomena spanning the wide spectrum of length and time scales. This cannot be accomplished with one theory or simulation method. Finally, we describe some of the emerging applications of virtual reality coupled to molecular dynamics simulations recently completed or currently underway in the field of bio-nanotechnology with envisioned applications into molecular robotics.

In Chap. 2, we investigated computational nanotechnology methodologies, i.e., physics-, biology- and chemistry-based modeling and simulation, dedicated to the design and optimization of bio-nano structures. An attempt has been made to emphasize the paradigm of how computational nanotechnology can be used to not only understand and characterize the systems under experimental investigation, but also as to how predictions can be made for novel applications because there is a strong coupling between the structural, mechanical, and biological properties at multiple length-time-scale. Two prototyping methodologies, namely, *interactive multiscale design* using virtual reality techniques and *co-prototyping design* using experimental measurements and molecular modeling algorithms are proposed and discussed in detail.

Then, Chap. 3 presents the design hypothesis, architectures and computational analysis of novel bio-nanorobotic structures. Towards this end, we used an interactive prototyping platform using multiscale, multiphysics and virtual reality advanced techniques. Virtual Reality technology is applied here, which not only provides immersive visualization but also gives an added functionality of CAD-based design, simulation, navigation and interactive manipulation of molecular graphical objects. The present study demonstrates that the mechanical properties of protein-based nanodevices can be controlled by using rational design based on protein and/or DNA engineering principles. We proposed novel concepts of protein-based and DNA-based actuation nanomechanisms that have been computationally simulated and analyzed through multiscale modeling. These designs pave the way for future position and force control nano servomotors operating in water environments.

Finally, Chap. 4 introduces a new co-prototyping approach based on a dual paradigm, i.e., experimentation and molecular modeling. Usually, nanodevice optimization is only based on experimental investigation of nanodevice characteristics. As all design parameters are not accessible by experimental measurements, the designer is not able to characterize deeply the nanodevice operating principles leading to optimal performances. As alternative, we propose to use jointly modeling tools to support the designer in order to characterize novel phenomenons, evaluate energy densities or new operational principles. Such principles have been applied to high-density electromechanical switches based on bi-directional linear bearing. Computational models provide new insights into the structural and energetic stability aspects of these unique nanomechanical elements.

Chapter 1
Current State-of-the-Art on Nanorobotic Components and Design

1 Introduction

This chapter focuses on the state of the art in the field of nano-robotics by describing various molecular level systems and associated design and characterization issues. Nano-robots are controllable machines at the nanometer (10^{-9}) or molecular scale that are composed of nano-scale components. With the modern scientific capabilities, it has become possible to attempt the creation of nanorobotic devices and interface them with the macro world for control. There are countless such machines that exist in nature and there is an opportunity to build more of them by mimicking nature. Even if the field of nanorobotics is fundamentally different than that of macro robots due to the differences in scale and material, there are many similarities in design and control techniques that eventually could be projected and applied. A roadmap towards the progression of this field is proposed and some design concept and philosophies are illustrated. There are many applications for nanorobotic systems and its biggest impact would be in the area of medicine.

Nanorobots can be defined as devices that can produce useful work through the interaction of individual molecules at the molecular scale of length. A convenient unit of measurement at the molecular scale would be a nanometer. Hence, molecular machines also fall into the category of nanomachines. Molecular machines depend on inter- and intramolecular interactions for their function. These interactions include forces such as the ionic and van der Waals forces and are a function of the geometry of the individual molecules. The interaction between two given molecules can be well understood by a set of laws governing them, which brings in a definite level of predictability and controllability of the underlying mechanics. Mother Nature has her own set of molecular machines that have been working for centuries and have become optimized for performance and design over the ages. As our knowledge and understanding of these numerous machines continues to increase, we now see a possibility of using the natural machines, or creating synthetic ones from scratch, by mimicking nature. In this review, we try to understand the principles, theory, and utility of the known molecular machines and look into the design and control issues for creation and modification of such machines. A majority of

M. Hamdi, A. Ferreira, *Design, Modeling and Characterization of Bio-Nanorobotic Systems*,
DOI 10.1007/978-90-481-3180-8_1,

natural molecular machines are protein based, whereas the DNA-based molecular machines are mostly synthetic. Nature deploys proteins to perform various cellular tasks, from moving cargo to catalyzing reactions, whereas DNA has been retained as an information carrier. Hence, it is understandable that most of the natural machinery is built from proteins. However, with the technological progress, it becomes possible to modify natural nanorobot structure, and even it proven possible to design a new nanorobotic devices concept that can be used for specific tasks such as drug delivery [1–3] or modern engineering actuation techniques [4–6]. The best nanomaterials candidates for this type of devices are the carbon nanotubes for their mechanics and electric properties, the proteins and DNA which are the suitable elements for the biomedical applications. The design of theses nanodevices require the characterization of candidates nanomaterials. In this chapter we will present the various molecular elements which can compose a nanorobot structure, such as the carbon nanotubes, the protein based elastic elements, biological actuators, inorganical actuator, carbon nanotube based actuator, as well as the existing methods of design and characterization of nanorobotic devices.

2 Nanorobotics Device Structures

Bio-nanorobotics is a new and rapidly growing interdisciplinary field addressing the assembly, construction and utilization of biomolecular devices based on nanoscale principles and/or dimensions. A key application is for medical target identification in therapeutical diagnosis, medical therapies and minimally invasive surgery (MIS) (Fig. 1.1) [1–3]. Modern engineering actuation techniques inspired by nature have been successfully implemented in microrobots evolving in fluidic environments using external electromagnetic fields [7–9]. On contrary, biological approach shows that flagellated propulsion mechanisms of bacteria [10], DNA-based conformation devices [11] or magnetic stereotaxis systems [12] represent a fertile territory for untethered nanoscale machines without the need of external hardware for actuation. In this way proteins could act as motors, mechanical joints, transmission elements, or sensors [6]. Successful implementation of bio-nanorobotics require the merging of both research approaches in order to address nanorobot choices on sensing, hardware architecture design, manufacturing approaches, and control methodology investigation [14]. However, when size goes down to nanoscale new physical properties and coupling effects between mechanical motion and physical, biochemical properties may become dominant, and multiphysics-scale problems arise. Additionally, the experimental techniques usually employed for motor proteins characterization have one limitation consisting in the direct observation occurring during mechanical tests. Due to the small scale structure of the proteins it is difficult to anchor them, assemble them, and control their behavior during the pulling experiments.

Proteins, DNA and carbon nanotube, regarding their biological proprieties and controllability characteristics can be the suitable elements for nanorobotics system for different applications. We are interested by the design and characterization of

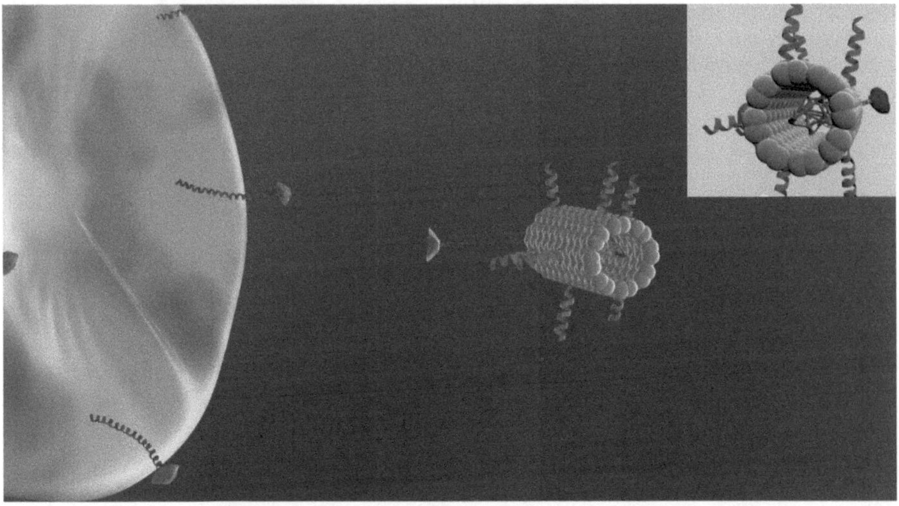

Fig. 1.1 Drug delivery nanorobotic system for the propulsion and navigation in the cardiovascular system through the induction of force from magnetic gradients generated by a clinical Magnetic Resonance Imaging (MRI)

such nanodevices by using computational methods coupled to experimental feedback. Design of nanorobotics system require the studies and characterization of the different nanomaterial candidates to be used for nanorobot fabrication. In this section we will present the various molecular elements used for nanorobotic device design.

Carbon Nanotube

Since their discovery in 1991 by Iijima and coworkers [15] carbon nanotubes have been investigated by many researchers all over the world. Their large length (up to several microns) and small diameter (a few nanometers) result in a large aspect ratio. They can be seen as the nearly one-dimensional form of fullerenes. Therefore, these materials are expected to possess additional interesting electronic, mechanic and molecular properties. Especially in the beginning, all theoretical studies on carbon nanotubes focused on the influence of the nearly one-dimensional structure on molecular and electronic properties. It is relatively easy to imagine a single-wall carbon nanotube (SWNT). Ideally, it is enough to consider a perfect graphene sheet (graphene is a polyaromatic monoatomic layer consisting of sp2-hybridized carbon atoms arranged in hexagons; genuine graphite consists of layers of this graphene) and to roll it into a cylinder (Fig. 1.2), making sure that the hexagonal rings placed in contact join coherently. Then the tips of the tube are sealed by two caps, each cap being a hemi-fullerene of the appropriate diameter (Fig. 1.3).

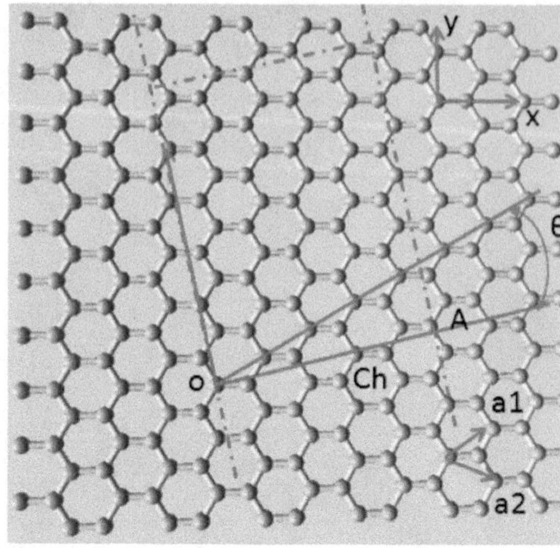

Fig. 1.2 Sketch of the way to make a single-wall carbon nanotube, starting from a graphene sheet

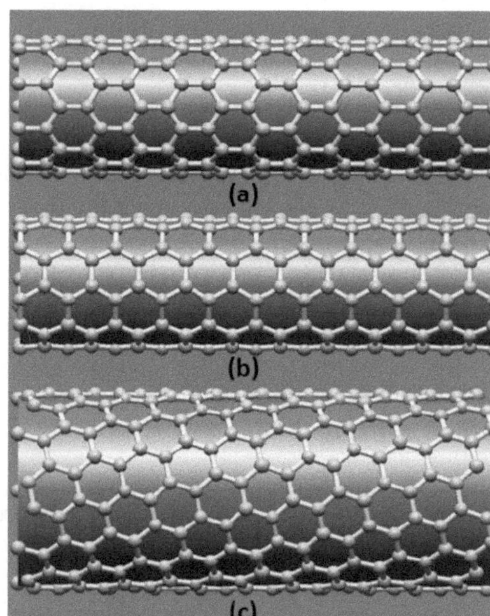

Fig. 1.3 Sketches of three different SWNT structures that are examples of (**a**) a zig-zag-type nanotube, (**b**) an armchair-type nanotube, (**c**) a helical nanotube

Geometrically, there is no restriction on the tube diameter. However, calculations have shown that collapsing the single-wall tube into a flattened two-layer ribbon is energetically more favorable than maintaining the tubular morphology beyond a diameter value of ≈2.5 nm [18]. On the other hand, it is easy to grasp intuitively that the shorter the radius of curvature, the higher the stress and the energetic cost, although SWNTs with diameters as low as 0.4 nm have been synthesized successfully [19]. A suitable energetic compromise is therefore reached for ≈1.4 nm, the most frequent diameter encountered regardless of the synthesis technique (at least for those based on solid carbon sources) when conditions ensuring high SWNT yields are used. There is no such restriction on the nanotube length, which only depends on the limitations of the preparation method and the specific conditions used for the synthesis (thermal gradients, residence time, and so on). Experimental data are consistent with these statements, since SWNTs wider than 2.5 nm are only rarely reported in the literature, whatever the preparation method, while the length of the SWNTs can be in the micrometer or the millimeter range. These features make single-wall carbon nanotubes a unique example of single molecules with huge aspect ratios. Two important consequences derive from the SWNT structure as described above:

1. All carbon atoms are involved in hexagonal aromatic rings only and are therefore in equivalent positions, except at each nanotube tip, where $6 * 5 = 30$ atoms are involved in pentagonal rings (considering that adjacent pentagons are unlikely)-though not more, not less, as a consequence of Eulers rule that also governs the fullerene structure. For ideal SWNTs, chemical reactivity will therefore be highly favored at the tube tips, at the locations of the pentagonal rings.
2. Although carbon atoms are involved in aromatic rings, the C=C bond angles are not planar. This means that the hybridization of carbon atoms is not pure sp2; it has some degree of the sp3 character, in a proportion that increases as the tube radius of curvature decreases. The effect is the same as for the C60 fullerene molecules, whose radius of curvature is 0.35 nm, and whose bonds therefore have 10% sp3 character [20].

As illustrated by Fig. 1.3, there are many ways to roll a graphene into a single-wall nanotube, with some of the resulting nanotubes possessing planes of symmetry both parallel and perpendicular to the nanotube axis (such as the SWNTs from Fig. 1.3(a) and Fig. 1.3(b), while others do not (such as the SWNT from Fig. 1.3(c)). Similar to the terms used for molecules, the latter are commonly called chiral nanotubes, since they are unable to be superimposed on their own image in a mirror. Helical is however sometimes preferred. The various ways to roll graphene into tubes are therefore mathematically defined by the vector of helicity C_h, and the angle of helicity θ as shown in Fig. 1.2.

$$OA = C_h = na_1 + ma_2 \tag{1.1}$$

with

$$a_1 = \frac{a\sqrt{3}}{2}x + \frac{a}{2}y \quad \text{and} \quad a_2 = \frac{a\sqrt{3}}{2}x - \frac{a}{2}y \tag{1.2}$$

where $a = 2.46$ Å and

$$\cos\theta = \frac{2n + m}{2\sqrt{n^2 + m^2 + nm}} \tag{1.3}$$

where n and m are the integers of the vector $\overrightarrow{OA}$ considering the unit vectors a_1 and a_2.

The vector of helicity C_h $(= OA)$ is perpendicular to the tube axis, while the angle of helicity θ is taken with respect to the so-called zig-zag axis: the vector of helicity that results in nanotubes of the zig-zag type (see below). The diameter D of the corresponding nanotube is related to C_h by the relation:

$$D = \frac{C_h}{\pi} = \frac{a_{cc}\sqrt{3(n^2 + 3m^2 + nm)}}{\pi}. \tag{1.4}$$

The C–C bond length is actually elongated by the curvature imposed by the structure; the average bond length in the C60 fullerene molecule is a reasonable upper limit, while the bond length in flat graphene in genuine graphite is the lower limit (corresponding to an infinite radius of curvature). Since C_h, θ, and D are all expressed as a function of the integers n and m, they are sufficient to define any particular SWNT by denoting them (n, m). The values of n and m for a given SWNT can be simply obtained by counting the number of hexagons that separate the extremities of the C_h vector following the unit vector a_1 first and then a_2 [16]. In the example of Fig. 1.2, the SWNT that is obtained by rolling the graphene so that the two shaded aromatic cycles can be superimposed exactly is a (4, 2) chiral nanotube. Similarly, SWNTs from Fig. 1.3(a) to Fig. 1.3(c) are (9, 0), (5, 5), and (10, 5) nanotubes respectively, thereby providing examples of zig-zag-type SWNT (with an angle of helicity $= 0°$), armchair-type SWNT (with an angle of helicity of 30) and a chiral SWNT, respectively. This also illustrates why the term chiral is sometimes inappropriate and should preferably be replaced with helical. Armchair (n, n) nanotubes, although definitely achiral from the standpoint of symmetry, exhibit a nonzero chiral angle. Zig-zag and armchair qualifications for a chiral nanotubes refer to the way that the carbon atoms are displayed at the edge of the nanotube cross-section (Fig. 1.3(a) and Fig. 1.3(b)). Generally speaking, it is clear from Fig. 1.2 and Fig. 1.3(a) that having the vector of helicity perpendicular to any of the three overall C=C bond directions will provide zig-zag-type SWNTs, denoted $(n, 0)$, while having the vector of helicity parallel to one of the three C=C bond directions will provide armchairtypeSWNTs, denoted (n, n). On the other hand, because of the sixfold symmetry of the graphene sheet, the angle of helicity θ for the chiral (n, m) nanotubes is such that $0 < \theta < 30°$. Figure 1.5 provides two examples of what chiral SWNTs look like, as seen via atomic force microscopy.

Multi Walled Nanotubes (MWNT) can be considered as a collection of concentric SWNTs with different diameters. The length and diameter of these structures differ a lot from those of SWNTs and, of course, their properties are also very different Fig. 1.4.

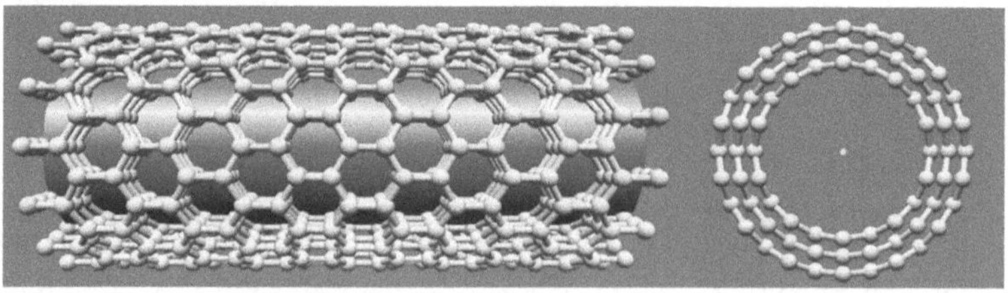

Fig. 1.4 Longitudinal and right view of a concentric multiwall carbon nanotube

Fig. 1.5 Image of two neighboring chiral SWNTs within a SWNT bundle as seen using high-resolution scanning tunneling microscopy (courtesy of Prof. Yazdani, University of Illinois at Urbana, USA)

Protein Springs as Passive Joints

Biological mechanisms and joints with multi-degree of freedom are currently investigated in the scientific community. As potential transmission components in bionanorobotics, we can note 2-dof molecular hinges based calix[4] arenes [13], DNA-based [11], linear ratchets mechanisms [23], molecular nanosprings using fibronectin and titin [21, 22] which can act as compliant joints in biomolecular robotic systems. Spring-like proteins are particularly interesting from their elastic, resilient and stability characteristics: stretch and relax without any net energy dissipation, can store and release energy and rectify motion in physical nanomachines in a full reversible way [12]. Protein flexibility refers to the capacity of the protein to experience dynamic changes in conformation under biological conditions.

In this thesis we are studying the development of protein based joint mechanisms. In order to draw analogies between mechanical protein methodologies and robotics methodologies, several protein-like elastic joints for structural links have been carefully studied and simulated to understand their functional and elastic limits. Contribution from molecular dynamic (MD) simulations is important in order to be able to understand the bio-nanomechanics of proteins and develop dynamic and kinematic models to study their performances. Computational studies of different classes of proteins: α-helix bundle proteins class (rop, DNA), β-sheet class (fibronectin, titin,) and compound class (collagen, silk, actin) have been carried out. It allows to predict the type of force spectra, reversibility, degrees of freedom and irreversible work that may be expected from molecular springs. Finally, quantitative models of biological spring mechanisms are developed and classified following their nanomechanical performances.

α-helix Bundle Proteins

Alpha-helices are the commonest secondary structural elements in a folded polypeptide chain, possibly because they are generated by local hydrogen bonding between C=O and N–H groups close together in the sequence. In an alpha helix, the carbonyl oxygen atom of each residue (n) accepts a hydrogen bond from the amide nitrogen four residues further along ($n + 4$) in the sequence Fig. 1.6(c), so that all of the polar amide groups in the helix are hydrogen bonded to one another except for the N–H group of the first residue in the helical segment (the aminoterminal

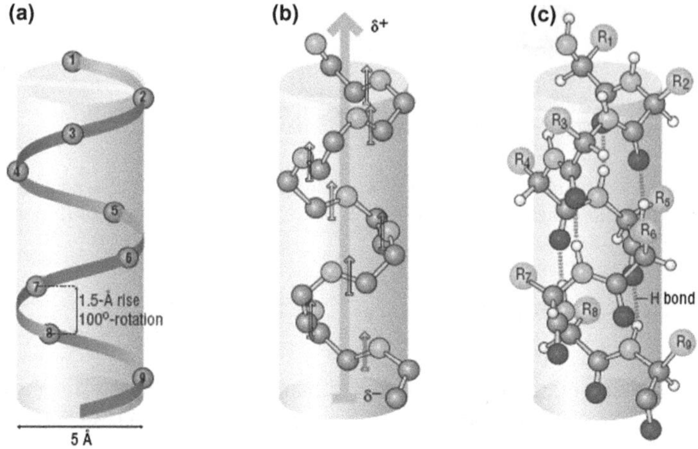

Fig. 1.6 (Color online) The alpha-helix. The chain path with average helical parameters is indicated showing (**a**) the alpha carbons only, (**b**) the backbone fold with peptide dipoles and (**c**) the full structure with backbone hydrogen bonds in *red*. All three chains run from top to bottom (that is, the amino-terminal end is at the top). Note that the individual peptide dipoles align to produce a macrodipole with its positive end at the amino-terminal end of the helix. Note also that the amino-terminal end has unsatisfied hydrogen-bond donors (N–H groups) whereas the carboxy-terminal end has unsatisfied hydrogen-bond acceptors (C=O groups). Usually a polar side chain is found at the end of the helix, making hydrogen bonds to these donors and acceptors; such a residue is called a helix cap (adapted from [36])

end) and the C=O group of the last one (the carboxy-terminal end). The result is a cylindrical structure where the wall of the cylinder is formed by the hydrogen-bonded backbone, and the outside is studded with side chains. The protruding side chains determine the interactions of the alpha helix both with other parts of a folded protein chain and with other protein molecules. The alpha helix is a compact structure, with approximate phi, psi values of −60° and −50° respectively: the distance between successive residues along the helical axis (translational rise) is only 1.5 Å (Fig. 1.6(a)). It would take a helix 20 residues long to span a distance of 30 Å. In a randomly coiled polypeptide chain the dipole moments of the individual backbone amide groups point in random directions, but in an alpha helix the hydrogen-bonding pattern causes all of the amides and their dipole moments to point in the same direction, roughly parallel to the helical axis (Fig. 1.6(b)). It is thought that, as a result, the individual peptide dipoles in a helix add to make a macrodipole with the amino-terminal end of the helix polarized positively and the carboxy-terminal end polarized negatively. The magnitude of this helix dipole should increase with increasing length of the helix, provided the cylinder remains straight. Because favorable electrostatic interactions could be made between oppositely charged species and the ends of the helix dipole, one might expect to find, at frequencies greater than predicted by chance, negatively charged side chains and bound anions at the amino-terminal ends of helices, and positively charged side chains and cations interacting with the carboxyterminal ends. Experimentally determined protein structures and studies of model peptides are in accord with these predictions. Indeed, the he-

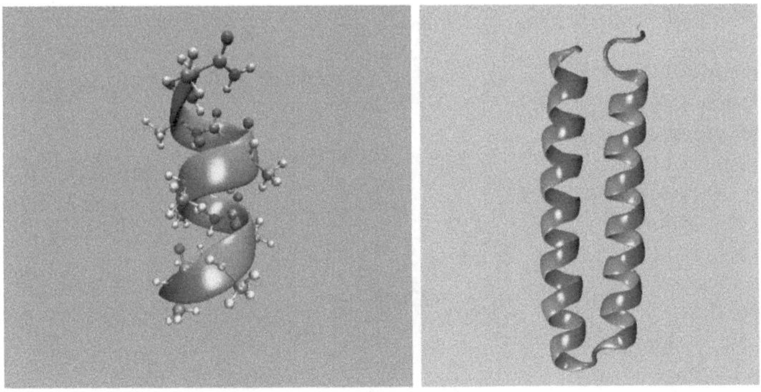

(a) (b)

Fig. 1.7 (**a**) Deca-alanine protein. (**b**) Helix bundle repressor of primer protein (ROP)

lix dipole in some cases contributes significantly to the binding of small charged molecules by protein.

1. *Alpha-alanine protein*: Alpha-alanine (Fig. 1.7(a)) is one of the nonessential amino acids, found in most proteins and particularly abundant in fibroin, the protein in silk. It is used in research and as a dietary supplement. Beta-alanine is a naturally occurring amino acid not found in proteins. It is an important constituent of the vitamin pantothenic acid and is used in its synthesis, as well as in biochemical research, electroplating, and organic synthesis. The low reactivity of the amino acid permits silk, a protein which contains some 30% alanine, to have a simple, elongated structure with few cross-links. This contributes to the desirable features of the fiber-strength, resistance to stretching, and flexibility.
2. *Repressor protein*: The repressor of primer protein (ROP) is a small, dimeric molecule consisting of two or four identical chains of 63 amino acids (Fig. 1.7(b)). Each monomer consists of two α-helices connected by a short turn and a seven-residue C-terminal tail. The two monomers pack together as a fully antiparallel four helixbundle. The bend region of ROP has attracted considerable interest as a parallel molecular spring due to its stability and elasticity properties [93].

Coiled-coil Alpha-helical Proteins

1. *Double-stranded helix myosin*: A particularly well characterized example of complementarity between interacting surfaces occurs in the case of coiled-coil structures (Fig. 1.8). Coiled coils are dimers of alpha helices formed through the ridges and grooves arrangement. We have already mentioned as the basis for tertiary structural interactions between alpha helices. In such interacting helices, hydrophobic side chains, often those of leucines, are repeated at intervals of seven amino acids in the chain, forming the ridge of hydrophobic side chains that fit into spaces on the interacting helix. This pattern is known as the heptad

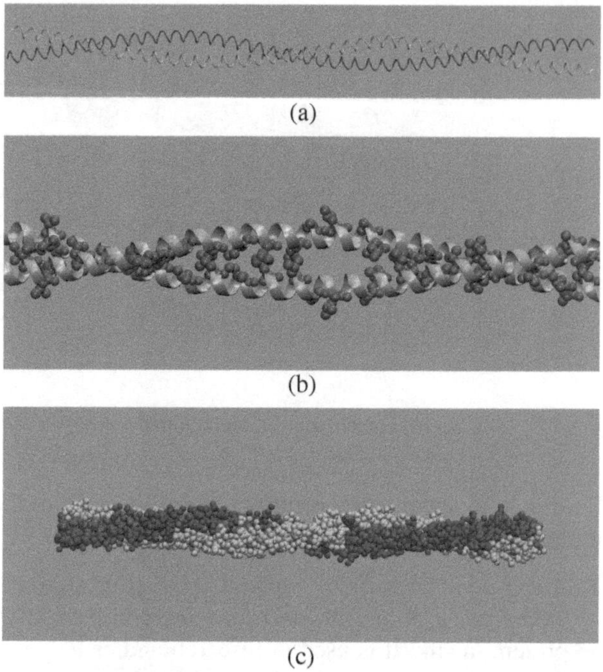

Fig. 1.8 Coiled-coil alpha-helical interactions: (**a**) Two interacting alpha helices of tropomyosin shown in a chain representation; (**b**) a space-filling representation of the separate alpha helices of tropomyosin with the hydrophobic side chains shown as dark protrusions; (**c**) the tropomyosin dimer showing how the hydrophobic side chains interdigitate in the coiled coil in a knobs in holes arrangement

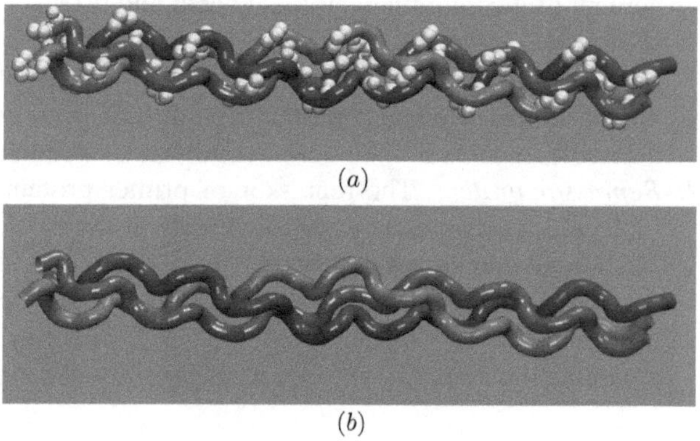

Fig. 1.9 The structure of collagen. Collagen is a three-chain fibrous protein in which each chain winds round the others. The rise per residue is much larger than in an alpha helix. (**a**) Three interacting alpha helices of tropocollagen; (**b**) representation with the hydrophobic side chains

repeat, and is characteristic of all dimeric structures formed through interacting alpha helices. It is one of the few cases in which structure can reliably be predicted from sequence.

2. *Triple-stranded helix tropocollagen protein*: Although the amino acid proline, which lacks an N–H group, is not frequently found in an alpha helix, two interesting helical structures can be formed from sequences rich in proline residues. The first is the collagen triple helix (Fig. 1.9). Collagen is the main constituent of the bones, tendons, ligaments and blood vessels of higher organisms and consists of a repeating tripeptide in which every third residue is a glycine (GlyXY)n. X and Y are usually proline residues, although lysine occurs sometimes. Many of

the proline residues are hydroxylated post-translationally. Each collagen strand forms a (left-handed) helical conformation and three such strands coil around each other like those of a rope. The effect is to create a fibrous protein of great tensile strength. Collagen molecules more than 2 μm in length have been observed. Denaturing the collagen triple helix by heating converts it to a disordered, dissociated, random mass that it is called gelatin.

β-sheet Proteins

In contrast to the alpha helix, the beta pleated sheet, whose name derives from the corrugated appearance of the extended polypeptide chain (Fig. 1.10), involves hydrogen bonds between backbone groups from residues distant from each other in the linear sequence. In beta sheets, two or more strands that may be widely separated in the protein sequence are arranged side by side, with hydrogen bonds between the strands (Fig. 1.10). The strands can run in the same direction (parallel beta sheet) or antiparallel to one another; mixed sheets with both parallel and antiparallel strands are also possible (Fig. 1.10).

Nearly all polar amide groups are hydrogen bonded to one another in a beta-sheet structure, except for the N–H and C=O groups on the outer sides of the two edge strands. Edge strands may make hydrogen bonds in any of several ways. They may simply make hydrogen bonds to water, if they are exposed to solvent; or they may pack against polar side chains in, for example, a neighboring alpha helix; or they may make hydrogen bonds to an edge strand in another protein chain, forming an extended beta structure that spans more than one subunit and thereby stabilizes quaternary structure (Fig. 1.11). Or the sheet may curve round on itself to form a barrel structure, with the two edge strands hydrogen bonding to one another to complete the closed cylinder (Fig. 1.12). Such beta barrels are a common feature of protein

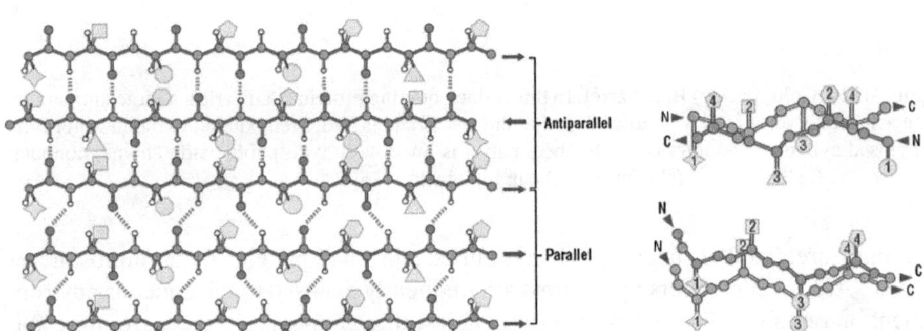

Fig. 1.10 The structure of the beta sheet: *The left figure* shows a mixed beta sheet, that is one containing both parallel and antiparallel segments. Note that the hydrogen bonds are more linear in the antiparallel sheet. On *the right* are edge-on views of antiparallel (*top*) and parallel sheets (*bottom*). The corrugated appearance gives rise to the name pleated sheet for these elements of secondary structure. Consecutive side chains, indicated here as numbered geometric symbols, point from alternate faces of both types of sheet (adapted from [36])

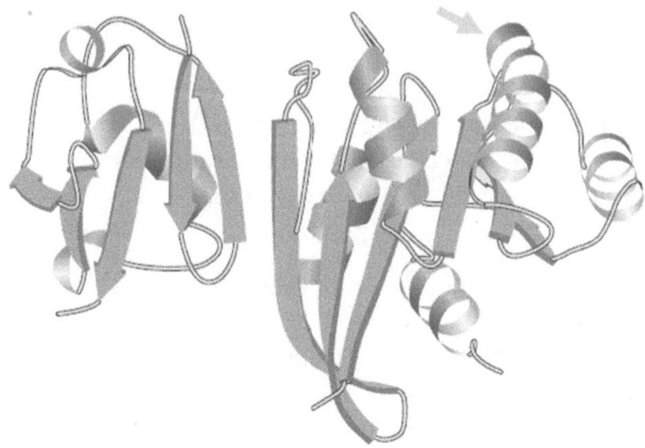

Fig. 1.11 (Color online) Two proteins that form a complex through hydrogen bonding between beta strands (the Rap-Raf complex, PDB 1gua). Two antiparallel edge strands of individual beta sheets hydrogen bond to each other at the protein-protein interface, forming a continuous mixed sheet that stabilizes the complex. The protein on the right contains a parallel beta sheet where each strand is connected to the next by an alpha helix, such as the one indicated with the *yellow arrow*. These helices pack against the faces of the sheet (adapted from [36])

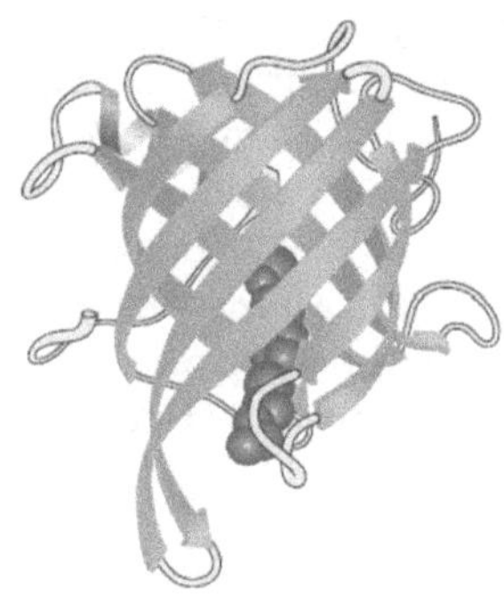

Fig. 1.12 (Color online) Beta barrel. In this retinol-binding protein (PDB 1rlb), a large antiparallel beta sheet curves all the way around so that the last strand is hydrogen bonded to the first, forming a closed cylinder. The interior of this beta barrel is lined with hydrophobic side chains; nonpolar molecules such as retinol (shown in *red*) can bind inside [36]

architecture. Parallel sheets are always buried and small parallel sheets almost never occur. Antiparallel sheets by contrast are frequently exposed to the aqueous environment on one face. These observations suggest that antiparallel sheets are more stable, which is consistent with their hydrogen bonds being more linear (see Fig. 1.10). Silk, which is notoriously strong, is made up of stacks of antiparallel beta sheets. We review in the following three β-sheets examples:

1. *Immunoglobulin-like (Ig) domains*: The giant muscle protein titin forms a filament which spans half of the sarcomere and performs, along its length, quite diverse functions. The region of titin located in the sarcomere I-band is be-

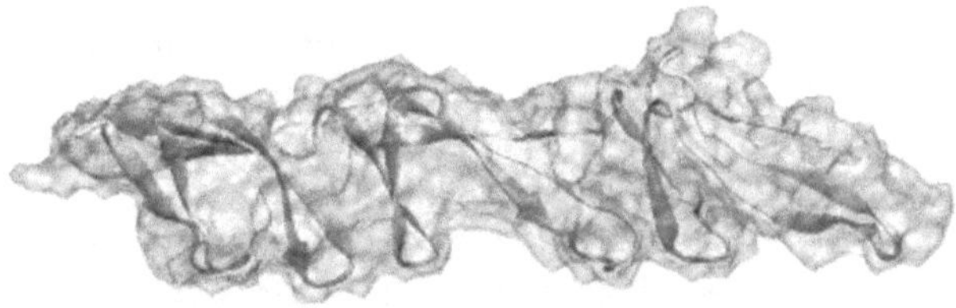

Fig. 1.13 (Color online) Fibronectin fragment. A cartoon of the secondary structure is displayed (beta-sheet in *yellow*, turns in *light blue*) within a transparent connolly surface

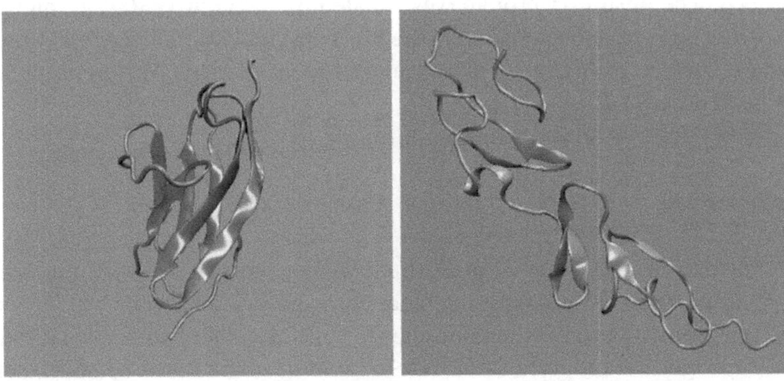

(*a*) (*b*)

Fig. 1.14 Immunoglobulin protein I27 model. Fibrillin protein model

lieved to play a major role in extensibility and passive elasticity of muscle. In the I-band, the titin sequence consists mostly of repetitive motifs of tandem immunoglobulin-like (Ig) modules intercalated by a potentially non-globular region. The highly repetitive titin architecture suggests that the molecular basis of its mechanical properties be approached through the characterization of the isolated components of the I-band and their interfaces. The immunoglobulin protein secondary structure consists of seven antiparallel β-strands stabilized primarily by hydrogen bonds between amide and carbonyl groups of the main chain. The I27 protein is certainly the best choice for biological nanospring (Fig. 1.14(a) due to its hyperelastic structure.

2. *Fibronectin*: Fibronectin (FN) is involved in many cellular processes, including tissue repair, embryogenesis, blood clotting, and cell migration/adhesion. Fibronectin exists in two main forms: (1) as an insoluble glycoprotein dimer that serves as a linker in the ECM (extracellular matrix), and; (2) as a soluble disulphide linked dimer found in the plasma (plasma FN). The plasma form is synthesized by hepatocytes, and the ECM form is made by fibroblasts, chondrocytes, endothelial cells, macrophages, as well as certain epithelial cells. Fibronectin sometimes serves as a general cell adhesion molecule by anchoring cells to collagen or proteoglycan substrates. FN also can serve to organize cellular interaction with the ECM by binding to different components of the extracellular matrix and to membrane-bound FN receptors on cell surfaces.

Fibronectin's structure is rod-like, composed of three different types of homologous, repeating modules, Types I, II, and III. These modules, though all part of the same amino acid chain, can be envisioned as "beads on a string," each one joined to its neighbors by short linkers.

3. *Fibrillin*: Fibrillin-1 is a major component of the microfibrils that form a sheath surrounding the amorphous elastin. It is believed that the microfibrils are composed of end-to-end polymers of fibrillin. To date, 3 forms of fibrillin have been described. The fibrillin-1 protein was isolated by Sakai in 1986 [24], and mutations in the gene have been linked to the Marfan syndrome. The general fibrillin's structure is show in Fig. 1.14(b). At present more than 100 different mutations have been described.

Deoxyribonucleic Acid (DNA) Acting as Passive Joint

DNA is not only in the focus of modern molecular biology, but also plays an increasingly important role as a building block for nanoscale materials and devices. In recent years, many researchers in nanoscience have used the unique, programmable molecular recognition properties of DNA to build nanostructures by self-assembly and to realize artificial, machine-like devices. We here give a brief survey of this field and discuss the possible applications of DNA-based nanodevices either as nanoscale motors and actuators, or as novel biosensors with built-in information-processing capability.

DNA molecule is the primary genetic material of most organisms. It is a double-helical biopolymer, Fig. 1.15, in which two chains of complementary nucleotides (the subunits whose sequence constitutes the genetic message) wind (usually right-handedly) around a common axis to form a double-helical structure [68]. Because of this unique structure, the elastic property of DNA molecule influences its biological functions greatly. There are mainly three kinds of deformations in DNA double-helix: stretching and bending of the molecule, twisting of one nucleotide chain relative to its counterpart. All these deformations have vital biological significance. During DNA replication, hydrogen bonds between the complementary DNA bases should be broken and the two nucleotide chains be separated. This strand separation process requires cooperative unwinding of the double-helix [69]. In DNA recombination reaction, RecA proteins polymerize along DNA template and the DNA molecule is stretched to 1:5 times its relaxed contour length [70, 71]. It is suspected that thermal fluctuations of DNA central axis might be very important for RecA polymerization [72]. Detailed study on DNA elasticity now becomes possible with the recent experimental developments, including, e.g., optical tweezer methods, atomic force microscopy, fluorescence microscopy. These techniques make it possible to manipulate directly single polymeric molecules and to record their elastic responses with high precision. Experiments done on double-stranded DNA (ds-DNA) have revealed that this molecule has very novel elastic property [76, 83]. When a torsionally relaxed DNA is pulled with a force less than 10 picoNewton (pN), its elastic response can be quantitatively understood by regarding the chain

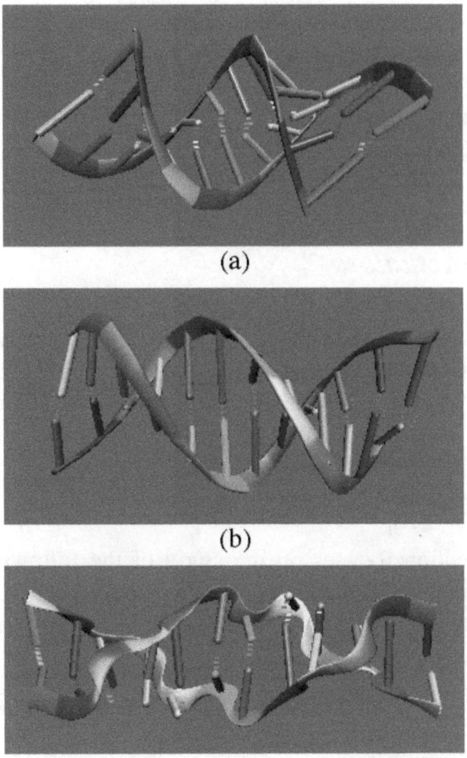

(a)

(b)

(c)

Fig. 1.15 DNA's structures: (**a**) A-DNA; (**b**) Z-DNA structures and (**c**) B-DNA

as an inextensible thin string with certain bending rigidity. However, if the external force is increased up to 65 pN, DNA chain becomes highly extensible. At this force, the molecule transit to an over-stretched configuration termed S-DNA, which is 1:6 times longer than the same molecule in its standard B-form structure [78, 79]. Besides external forces, it is also possible to apply torsional constraints to DNA double-helix by external torques. The linking number of DNA, i.e., the total topological turns one DNA strand winds around the other, can be fixed at a value larger (less) than the molecules relaxed value. In such cases we say the DNA molecule is positively (negatively) supercoiled. It is shown experimentally [80] that when external force is less than a threshold value of about 0:3 pN, the extension of DNA molecule decreases with increasing twisted stress and the elastic response of positively supercoiled DNA is similar to that of negatively supercoiled DNA, indicating the DNA chain might be regarded as a chiral. However, if the external force is increased to be larger than this threshold, negatively and positively supercoiled DNA molecules behave quite differently. Under the condition of fixed external force between 0:3 pN and 3 pN, while positive twist stress keeps shrinking the DNA polymer, the extension of negatively supercoiled DNA is insensitive to supercoiling degree [80, 81]. In higher force region, it is suggested by some authors that positively supercoiled DNA may transit to a configuration called Pauling-like DNA (P-DNA) with exposed nucleotide bases [82], while negative torque may lead to strand-separation in DNA molecule (denaturation of DNA double-helix [81]). A very recent systematic observation performed by Leger et al. [83], on the other hand, suggested another possibility that negative supercoiling may result in left-handed Z-form configuration in DNA (Fig. 1.15).

NanoActuators

In this section we will detail some of the man made and naturally occurring molecular machines. We divide the molecular machines into three broad categories—protein-based, DNA-based and chemical molecular motors.

Protein Based Nanomotor

This section focuses on the study of the following main protein based and DNA based molecular machines: ATP Synthase, the Kinesin, Myosin and Flagella Molecular Motors.

1. ATP-BASED PROTEIN nanomotors: Three naturally existing rotary motors have been identified and studied in detail so far. Two form the F0F1-ATP synthase, and the third one is the bacterial flagellar motor. The protein-based molecular motors rely on an energy-rich molecule known as adenosine triphosphate (ATP), which is basically a nucleotide having three phosphate molecules that play a vital role in its energetics, and make it an indispensable commodity of life. The machines described in this section, the F0F1-ATPase, the kinesin, myosin, and dynein [37] superfamily of protein molecular machines, and bacteria flagellar motors all depend, directly or indirectly, on ATP for their input energy. These machines, which have been carrying out vital life functions both inside and outside cells for millions of years, have now been segregated out of their natural environment and are seen as energy conversion devices to obtain forces, torques, and motion. One disadvantage associated with ATP dependence is that the ATP creation machinery itself could be many times heavier and bulkier than the motors, thereby making the assembly more complex. These machines perform best in their natural environment, and in the near future it may not be possible to have them as a part of feasible biomimetic molecular machinery [42].

   ATP is regarded as the energy currency of biological systems. The ATP molecule owes much of its energy to the terminal three phosphate ions attached to an adenosine base. In 1941 the role of ATP in the energy conversion process in living beings was recognized. However, the mode of transfer and structure of the enzymewas unknown. When this currency is utilized (i.e., the energy of the molecule that is used to drive a biological process), the terminal anhydride bond in the ATP molecule has to be split. This leaves adenosine diphosphate (ADP) and a phosphate ion (Pi) as the products, which are recombined to form ATP by a super efficient enzyme motor assembly called the F0F1-ATP synthase (F0F1-ATPase) [37, 38, 42]. ATP synthase is present inside the mitochondria of animal cells, in plant chloroplasts, in bacteria, and some other organisms. ATP synthase was first seen in 1962 in an electron microscopy experiment on bovine heart mitochondria, as 10 nm diameter knobs. Their importance in energy conversion was realized, but their functioning was still unknown. In 1966 the relation of the thus far unknown knobs to the production of ATP was established, which provided one of the first structures of the enzyme. The ATP synthase is actually a combination of two motors functioning together, the hydrophobic transmembrane F0-ATPase motor and the globular F1-ATPase motor [37, 38, 42]. Both motors have distinct structures and functions. There are different abbreviations used for the F1-ATPase based on their sources; the heart mitochondrial motors are called mF1, chloroplast motors are cF1, those obtained from Escherichia coli are termed EcF1, and the ones from Kagawa's thermophilic bacterium are known as TF1 [39]. The F0 motor has organism-dependant structural variations. In addition, the regulation of catalysis in ATP synthase depends on the organism's source. In animal mitochondria, this motor is embedded in the inner mitochondrial membrane and uses an ion-motive force for its function. Initially, however, it was believed that the force was proton-motive only until it was shown that, in some cases, Na^+ ions induce the motive force for the F0 motor; hence the term ion-motive force. The proton-motive force can be defined as the work per unit charge that a proton traveling through a membrane can perform [42].

The F1 motor, powered by hydrolysis of ATP, is composed of a central protein stalk, called the γ-subunit, surrounded by three copies each of α- and β-subunits. The α- and β-subunits are arranged alternately so that they make a symmetric circular pattern when viewed from the top. There are δ-subunits attached to the periphery of the α–β cylinder and the α-subunits are present at the base of the γ-subunits, as shown in Fig. 1.16 [37]. Hence, the F1 motor is composed of nine polypeptides. The α- and β-subunits contain nucleotide-binding sites that bind ATP/ADP molecules. The nucleotidebinding sites in the á-subunits simply bind the nucleotide, whereas those in the β-subunit actually perform the catalysis. The a, b, and the c subunits shown in Fig. 1.16 are a part of the F0 motor discussed below.

The binding-change mechanism to explain the function of F1-ATPase was proposed in 1973 [42]. The mechanism, as known today, shows that each of the â-subunits take three forms: O (open), L (loose), and T (tight) binding site. When the subunit is in the O form, it is catalytically inactive and has very low affinity to bind substrates. In the L form, the subunit loosely binds substrates (ADP and Pi), although it is catalytically inactive. In the T form, the ADP and Pi are converted into a tightly bound ATP until a conformational change converts the T-site into an O-site, thereby allowing the release of the newly formed ATP [40, 41]. The mechanism is shown in Fig. 1.17 [38]. The experimental setup is shown in Fig. 1.18 [39]. F1-ATPase can produce 80–100 pN-nm of rotary torque.

2. *The Myosin, Kinesin, and Flagella Molecular Motors*: With modern microscopic tools, we view a cell as a set of many different moving components powered

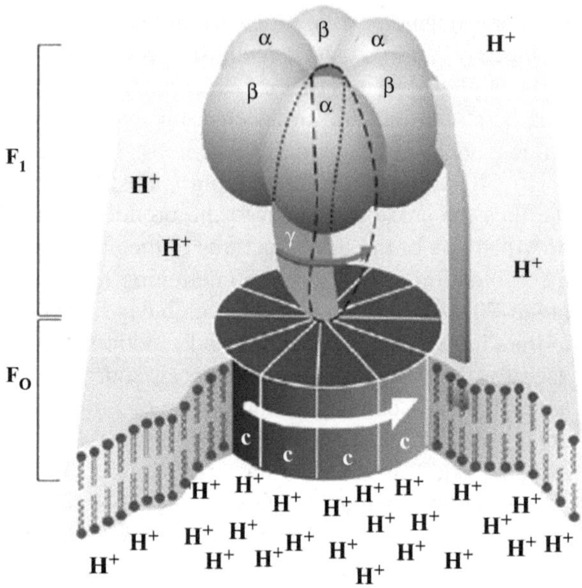

Fig. 1.16 The F0F1-ATPase motors. The F0 motor is embedded in the inner mitochondrial membrane of the mitochondria. F0 is typically composed of a, b, and c subunits as shown. The F1 motor is the soluble region composed of three α-, three β-, one each of γ-, α- and ϵ-subunits [37]

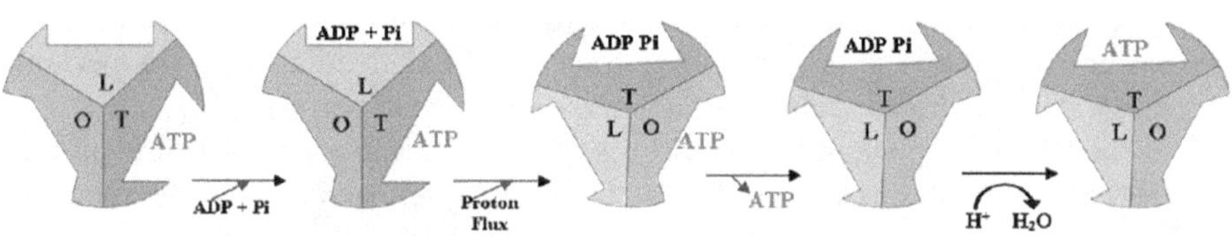

Fig. 1.17 The binding-change mechanism of F1-ATPase. The three catalytic sites bind ADP/ATP alternately in L (loose), T (tight) and O (open) fashion. ADP and Pi are initially loosely bound, then the binding becomes tight, with the conversion of ADP + Pi into ATP, which is finally released when the open conformation is achieved [38]

by molecular machines rather than a static environment. Molecular motors that move unidirectionally along protein polymers (actin or microtubules) drive the motions of muscles, as well as much smaller intracellular cargoes. In addition to the F0F1-ATPase motors inside the cell, there are linear transport motors present, tiny vehicles known as motor proteins, that transport molecular cargoes and also require ATP for functioning. These minute cellular machines exist in three families: kinesins, myosins, and dyneins. The cargoes can be organelles, lipids, or proteins, etc. They play an important role in cell division and motility. There are over 250 kinesin-like proteins, and they are involved in processes as diverse as themovement of chromosomes and the dynamics of cell membranes. The only part they have in common is the catalytic portion known as the motor domain. They have significant differences in their location within cells, their structural organization, and the movement they generate. Much structural information about kinesin is now available through the crystal structures. The motor domain contains a folding motif similar to that of myosin and G proteins (36). The two heads

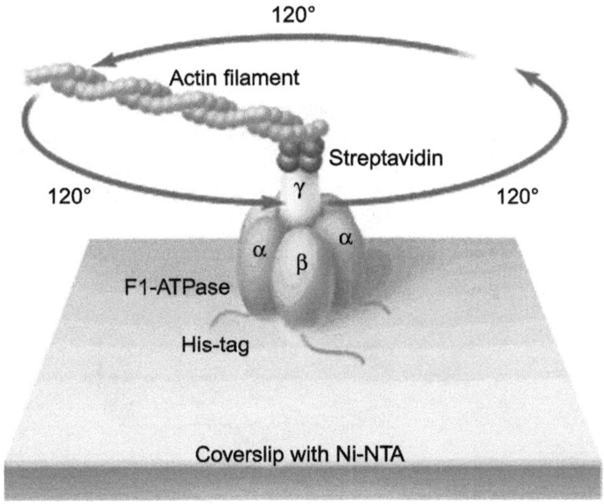

Fig. 1.18 Rotation of the ATPase motor as shown experimentally in [37]. A 1–4 μm long fluorescently tagged actin filament was attached to the F0-ATPase using streptavidin to observe the rotation of the ATP synthase motor [39]

or the motor domains of kinesin are linked via neck linkers to a long coiled coil, which extends up to the cargo (Fig. 1.19(b)). These heads interact with the α- and β-subunits of the tubulin hetrodimer along the microtubule protofilament. The heads contain nucleotide- and microtubule-binding domains. Although kinesin is also a two-headed linear motor, its modus operandi is different from myosin in the sense that both of its heads work together in a coordinated manner in contrast to one being left out in the case of myosin. Figure 1.19(b) shows the kinesin walk. Each of the motor heads is near the microtubule in the initial state, with each motor head carrying an ADP molecule. When one of the heads loosely binds to the microtubule, it loses its ADP molecule to facilitate a stronger binding. Another ATP molecule replaces the ADP, which facilitates a conformational change such that the neck region of the bound head snaps forward and zips on to the head, (courtesy C. Mavroidis, molecular machine).

3. *The flagella motors*: The flagellum is the main organelle for motility in bacteria. Despite bearing the same name, bacterial flagella are distinct in form, function and evolution from both archaeal and eukaryotic flagella. The archetypal bacterial flagellum from Salmonella enterica serovar Typhimurium (Salmonella typhimurium) consists of a basal body, embedded in the cell wall, and two axial structures, the hook and filament, which are joined at the hook-filament junction (Fig. 1.20) [41]. The basal body consists of the MS ring, rod, and L- and P-rings. Components of the axial structures are exported from the cell by the flagellar type III secretion system, which consists of several proteins from the MS ring and a peripheral hexameric ATPase FliI that drives the export process. Exported proteins pass through a central pore in the axial structures-therefore, first the flagellar hook and then the filament grow by the incorporation of new subunits at the distal end of the filamentous structure. Rotation of bacterial flagella is powered by a proton- or sodium-motive force. The flagellar motor converts electrochemical energy into torque through an interaction between two components:

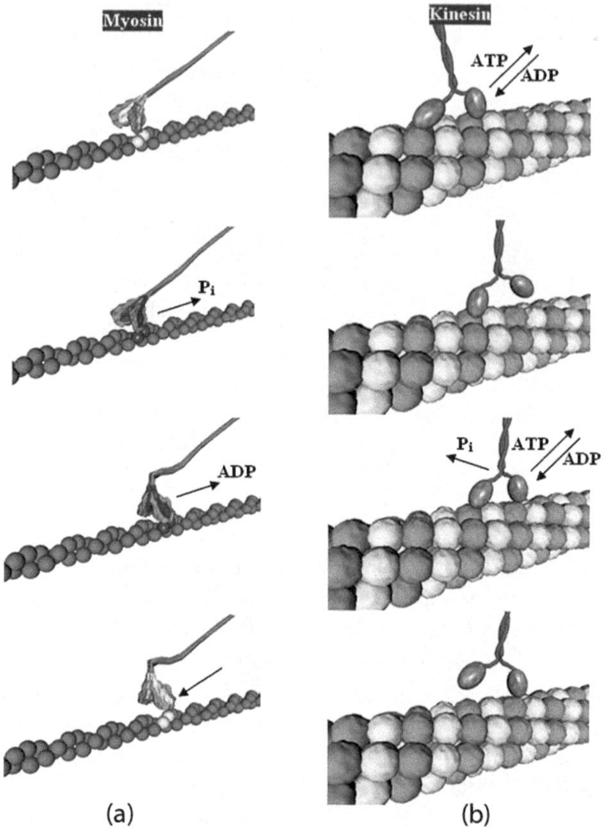

Fig. 1.19 (Color online) The kinesin-myosin walks: (**a**) Myosin motor mechanism. (i) Motor head loosely docking to the actin binding site; (ii) The binding becomes tighter along with the release of Pi; (iii) Lever arm swings to the left with the release of ADP, and; (iv) replacement of the lost ADP with a fresh ATP molecule results in dissociation of the head; (**b**) Kinesin heads working in conjunction. (i) Both ADP-carrying heads come near the microtubule and one of them (*black* neck) binds; (ii) Loss of bound ADP and addition of fresh ATP in the bound head moves the other (*red* neck) to the right; (iii) The second head (*red*) binds to microtubule while losing its ADP, and replacing it with a new ATP molecule while the first head hydrolyses its ATP and loses Pi; (iv) The ADP-carrying *black*-neck will now be snapped forward, and the cycle will be repeated [42]

the stator and the rotor. The stator consists of multiple copies of two proteins, MotA and MotB, which assemble into a structure that is associated with the inner membrane and anchored to peptidoglycan, so that it remains stationary. The rotor consists of multiple copies of FliG, which together with FliM and FliN form the C ring, mounted on the cytoplasmic face of the MS ring. Torque is transmitted from the C-ring by the MS ring to the rod (a molecular drive shaft) and from there to the hook (a universal joint) and then on to the helical flagellar filament (a molecular propeller). Rotation of this helical filament converts torque into thrust, conferring motility on the cell. The chemotaxis apparatus (not shown)

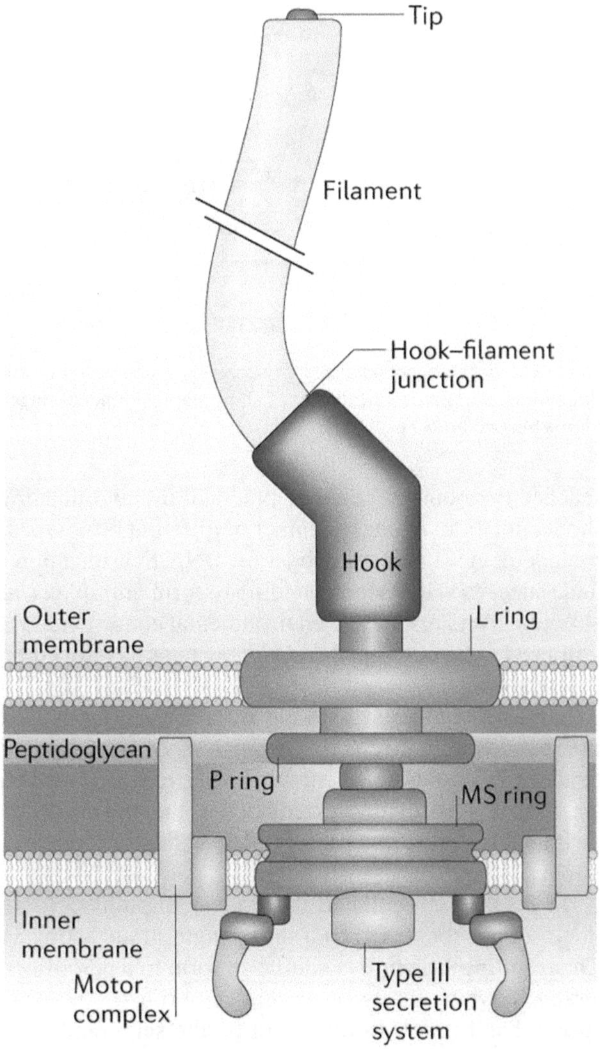

Fig. 1.20 A typical flagellum. A filament is connected to the hook, which connects to the transmembrane motor unit through a shaft. Hook-related proteins help in assembly and stability of the hook and filament. The L-ring is embedded in the outer cell membrane, the P-ring in the peptidoglycan layer, and the MS-ring along with rotor and parts of stator are embedded in the inner cell membrane. The C-ring and the transport apparatus are located inside the cell [41]

integrates diverse signals to modulate the behaviour of the motor so as to propel the cell towards nutrients. Several soluble factors are involved in coordinating the assembly of the flagellar apparatus, including the flagellar factor FliA and the hook-length control protein, FliK.

DNA-based Nanomotor

DNA-based nanodevices that can move or change conformation have been developed rapidly since the first example was reported in [109]. However, none have

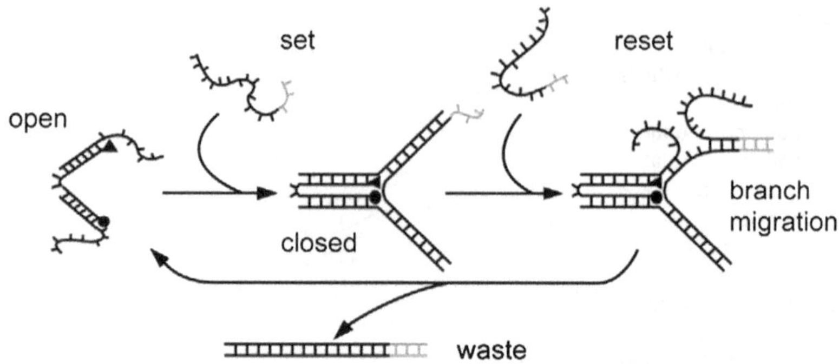

Fig. 1.21 In the open state DNA tweezers are composed of three strands of DNA forming two double-helical arms connected by a short single-stranded hinge. Hybridization with a set strand closes the two arms [110]

reached the point of being of practical use and thus toward is to be emphasized in the section title. So far, a number of different prototypes have been developed. These include devices that are driven by DNA hybridization and branch migration, walk and rotate, use Hoogsteen bonding to form multiplex (rather than double helix) DNA structures, and respond to environmental conditions. At least some of these concepts will need to be combined and/or new ones developed before such devices will be of practical use.

Three major features arising from Watson-Crick base pairing have been put into use: the different mechanical properties of ssDNA and dsDNA; single-stranded extensions (toeholds); and strand displacement (branch migration). For branch migration, single-stranded extensions at the end of dsDNA are used as points of attachment for effector strands that specifically remove single DNA strands from the duplex.

In 2000, [110] collated these ideas to build DNA tweezers in which two DNA duplexes are connected by a short single strand acting as a flexible hinge (Fig. 1.21). The resulting structure is similar in form to a pair of open tweezers. By adding a set strand to which the tweezers ends hybridize, the tweezers can be closed. A reset strand that attaches to a toehold on the set strand is then added. This reset strand displaces the set strand from the tweezers through branch migration, freeing them to open again. The tweezers' ends are labeled with a donor/acceptor fluorophore pair, and their relative movement is measured using FRET techniques

Using similar concepts, Simmel and Yurke [111] have developed a nanoactuator that switches from a relaxed, circular form to a stretched conformation. DNA scissors have been demonstrated by [112] in which two sets of tweezer structures are joined at their hinges with short carbon linkers. The motion of one set of tweezers is transduced to the other part, resulting in a scissor movement. Recently, a DNA-based nanodevice resembling tweezers has been applied as a force gauge to establish how much work a DNA-distorting protein can do when it binds to DNA [113].

The tweezers have been refined [114] who used the action of a DNAzyme (DNA molecule with enzymatic activity) to drive the device autonomously. Single molecule analysis of tweezers developed by [102] has led to the conclusion that addition of the closing strand does not always lead to perfect closure of the tweezers. Instead, other structures can occur, such as bonding of multiple closing strands per tweezer device, without the device closing. These unwanted forms need to be addressed with improved sequence-design of the device. Reference [115] have now developed fully addressable DNA-based molecular tweezers along a DNA track to actuate coupling reactions in a programmable fashion.

An autonomous device using ligase to bind connections, and restriction enzymes to cleave connections, has been developed by Yin et al. [120]. In contrast, Tian et al. [121] have produced DNAzyme walkers that move along a track consisting of RNA. Notably, the enzymatic properties of the walker provide cleaving as required, doing away with the need for extra restriction enzymes and ligases. A similar concept was recently used to construct molecular spiders [122].

As already mentioned, DNA can also form triplexes. In a DNA triplex, a third DNA strand binds along the major groove of a regular B-form double helix by Hoogsteen bonding, a base-pairing mode alternative to the Watson-Crick scheme. Chen et al. [114] varied the pH to form an ordered triplex helix from a duplex and a disordered strand, resulting in nanomechanical motion. This has been used to control chemical reactivity.

Inorganic Molecular Machines

1. *The Rotaxane*: Rotaxane family of molecular machines is characterized by two parts: a dumb-bell shaped compound with two heavy chemical groups at the ends and a light, cyclic component, called macrocycle, interlocked between the heads as shown in Fig. 1.22. It has been shown [126] that a reversible switch can be made with a rotaxane setup. For this, one needs to have two chemically active recognition sites in the neck region of the dumb-bell. In this particular example, the thread was made of polyether, marked by recognition sites hydroquinol units and terminated at the ends by large triisoproplylsilyl groups. A tetracationic bead was designed and self-assembled into the system that interacts with the recognition sites. The macrocycle has a natural low energy state on the first recognition site, but can be switched among the two sites reversibly upon application of suitable stimuli. Depending on the type of rotaxane setup the stimuli can

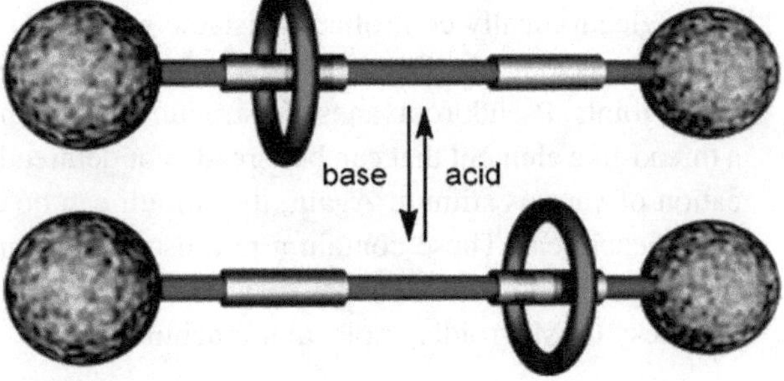

Fig. 1.22 A typical rotaxane shuttle set-up. The macrocycle encircles the thread-like portion of the dumb-bell with heavy groups at its ends. The thread has two recognition sites which can be altered reversibly so as to make the macrocycle shuttle between the two sites [126]

be chemical, electrochemical or photochemical [127, 128]. The stereo-electronic properties of the recognition sites can be altered protonation or deprotonation, or by oxidation or reduction, thereby changing their affinity towards the macrocycle. In a recent example, light induced acceleration of rotaxane motion was achieved by photoisomerization [129], while similar controls through alternating current (oscillating electric fields) was shown before [130].

2. *The Catenanes*: The catenanes are also special type of interlocked structures that represent a growing family of molecular machines. They are synthesized by supramolecular assistance to molecular synthesis [131]. The general structure of a catenane is that of two interlocked ring-like components that are non-covalently linked via a mechanical bond, that is, they are held together without any valence forces. Both the macrocyclic components have recognition sites that are atoms of groups that are redox-active or photochemically reactive. It is possible to have both rings with similar recognition sites. In such a scenario, one of the rings may rotate inside the other with the conformations stabilized by non-covalent interactions, but the two states of the inner ring differing by 180° will be undistinguishable (degenerate) [132]. For better control and distinguishable molecular conformations, it is desirable to have different recognition sites within the macrocycles. Then they can be controlled independently through their own specific stimuli. The stereo-electronic property of one recognition site within a macrocycle can be varied such that at one point it has more affinity to the sites on the other ring. At this instant, the force balance will guide the rotating macrocycle for a stable conformation that requires that particular site to be inside the other macrocycle. Similarly, with other stimulus, this affinity can be turned off, or even reversed along with the affinity of the second recognition site on the rotating macrocycle increased towards those on the static one. There is a need for computational modeling, simulation and analysis of such molecular machine motion [133]. Like rotaxanes, catenanes also can be designed for chemical, photochemical or electrochemical control. Figure 1.23 describes one such catenane molecular motor.

For both rotaxane and catenane based molecular machines, it is desirable to have recognition sites such that they can be easily controlled externally. Hence it is preferable to build sites that are either redox-active or photo-active. Catenanes can also be self assembled. An example of catenane assembled molecular motors is the electronically controllable bistable switch. An intuitive way of looking at catenanes is to think of them as molecular equivalents of ball and socket and universal joints. Pseudorotaxanes are structures that contain a ring-like element and a thread-like element that can be threaded or dethreaded onto the ring upon application of various stimuli. Again, the stimuli can be chemical, photochemical or electrochemical. These contain a promise of forming molecular machine components analogous to switches and nuts and bolts from the macroscopic world (courtesy C. Mavroidis, molecular machine).

Fig. 1.23 Redox controlled ring rotation in a catenane containing a non-symmetric ring [131]

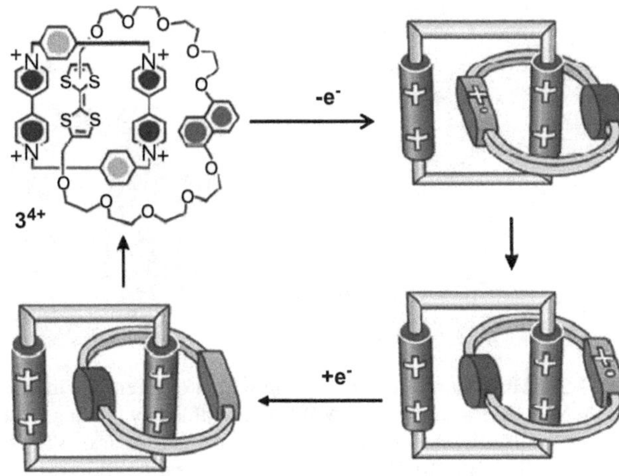

Carbon Nanotube Based Nano-motor

Since the discovery of carbon nanotubes [15] their unique electronic and mechanical properties are finding promising applications in nanoelectronics and nanomechanics. The weak van der Waals interaction between walls of a nanotube allows them easily to slide and rotate with respect to one another. The arbitrary [86, 87] and controlled by a manipulator [87] telescopic extension of walls in many-wall carbon nanotubes has been recently experimentally studied. This property provides a possibility to construct a new set of mechanical and electromechanical nanodevices where the nanometer size motion is the relative sliding, rotation or screwlike motion of nanotube walls. This set includes now several possible devices: nanobearings [88] and nanogears [89] driven by a laser electric field; constant-force nanosprings [87], mechanical nanoswitch [90], electrical nanoswitch [91], gigahertz oscillator [92] and nanodrill [91] were proposed. Based carbon nanotube Linear Servomotors and nanobearing, Fig. 1.24 was designed and fabricated by B.J. Nelson et al. [4, 5].

3 Virtual Reality Techniques for Bio-nanotechnology Design

Objective

Virtual Reality (VR) technology is applied here, which not only provides immersive visualization but also gives an added functionality of CAD-based design, simulation, navigation and interactive manipulation of molecular graphical objects. Although many of the described technologies have been developed into more or less mature products for robots acting in the macro-world, the nano-size of the molecular robots poses extreme challenges and requires a complete rethinking of the visual and haptic perception of nano-world. Using simulated biological nano environments in VR,

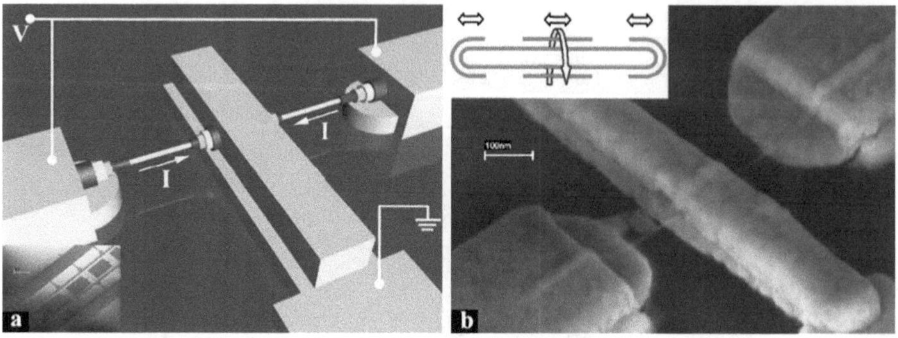

Fig. 1.24 Formation of a nanobearing with a different architecture via piecewise engineering of single NTs. (**a**) Illustration of a MWNT assembled by floating electrode DEP. The electrode wiring scheme for parallel shell structuring is also shown. *The inset* shows an SEM image of the fabricated array at a tilt of 40. (**b**) SEM image of a nanobearing formed by this method shown at a stage tilt of 40, with the scale bar representing 100 nm. *The inset* illustrates the degrees of freedom of the NT shell structure [5]

the operator can design and characterize through physical simulation the behavior of nano and bionano robots. Adding haptic interaction, the operator can explore and prevent the problems of bionano robots in their native environment. These immersive VR techniques allows the 3D visualization of scientific data and for modeling and animating complex nanoengineering systems. We describe in the following some of the emerging applications of VR recently completed or currently underway in the field of bio-nanotechnology with envisioned applications to molecular robotics.

Virtual Reality Techniques for Modeling and Simulation

Structural information on biological macromolecules is an essential requirement for our understanding of biological functions. From the very beginning of structural biology visualization was essential for determining and understanding structures. The initial work in this area was the use of Visualization and Virtual Reality for Bio-informatics, especially for the 3D structural analysis of bio-molecular system. Surface and volume based visualization provide three-dimensional concepts of biomolecular structure. Virtual Reality offers a channel to reach-into the molecular space in an immersive and interactive environment. Progress in the X-ray and nuclear magnetic resonance (NMR) instrumentation and computer and software technology has led to an increasing rate of accumulation of new biological structures such as proteins and DNA. The Protein Data Bank [43] has a database of coordinate entries of 21,248 proteins (Fig. 1.25). Proteins are complex biological structures made up of several hundred atoms. The usual approach to visualize such a molecule is to retrieve the coordinate files from the database and then use one of the molecular graphics software packages. These packages largely provide mono

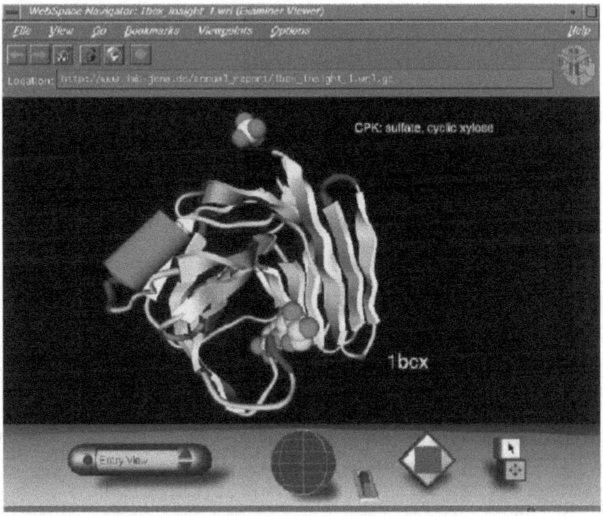

Fig. 1.25 Secondary structure representation of the Bacillus circulans xylanase (PDB code–1bcx) complexed with sulfate and cyclic xylose in the VRML viewer WebSpace (helix–tube, sheet–ribbon) (courtesy of: Dr. Jürgen Sühnel, Institut für Molekulare Biotechnologie, Germany)

representations of molecular images without any interactivity. Use of Virtual Reality Modeling Language (VRML) can provide a 3D and immersive environment with provision for stereo representations of the molecules [44]. This not only helps in better understanding the molecular structure but also provides insight into the chemical and bio-chemical properties of the molecule such as number of hydrogen bonds.

Complex Molecular Structure Modeling for Design and Analysis

The bio-systems are made up of atoms and since atoms cannot be observed directly, we can see them only in our imaginations. Commonly available physical models provide an intuitive representation of structural molecular biology. When applied to large molecules, such as proteins and the molecular assemblies found in cells, computer graphics simulation can be used to accurately portray various molecular computational models, with their varying complexities. Electrostatic field data around the molecules of interest, interatomic forces, laws of quantum mechanics, and so on represents the variety of bio-molecular nano-interactions [45]. However, these methods lack the beneficial tactile and kinaesthetic attributes of real physical models. By coupling an accurate molecular dynamics simulation code to an immersive virtual reality display with interactive capabilities and manual force feedback, "immersive" visualization of molecular atoms could be improved [46]. In order to create multi-modality enhancements of such tangible models, VR-based technology is an interesting tool. By superimposing additional graphical information on top of fabricated models, e.g. augmented reality, by incorporating support for voice commands and by providing haptic interface for sensing the electrostatic charges and interatomic collisions, the user would be able to interact with these virtual enhancements haptically while manipulating the physical model.

Molecular Dynamics Simulations in Virtual Environment

Molecular dynamics (MD) is a computer simulation technique where the time evolution of a set of interacting atoms is followed by integrating their equations of motion. In MD we follow the laws of classical mechanics, and most notably Newton's law. MD simulations are based on the calculation of the free energy that is released during the transition from native to fusogenic state and compute the atomic trajectories and coordinates by numerically solving the equations of motions using an empirical force field that mimics the actual inter-atomic forces in the molecular system. MD simulations of complex molecular systems require enormous computational power and produces large amount of data in each step. The resulting data includes number of atoms per unit volume, atomic positions, velocity of each atom, force applied on each atom and the energy contents. These results of MD simulations need to be visualized to give the user a more intuitive feel of what is happening. Haptic interaction used in conjunction with VR visualization helps the scientist to control/monitor the simulation progress and to get feedback from the simulation process as well [47]. Figure 1.26 shows the virtual reality visualization of molecular dynamics simulation.

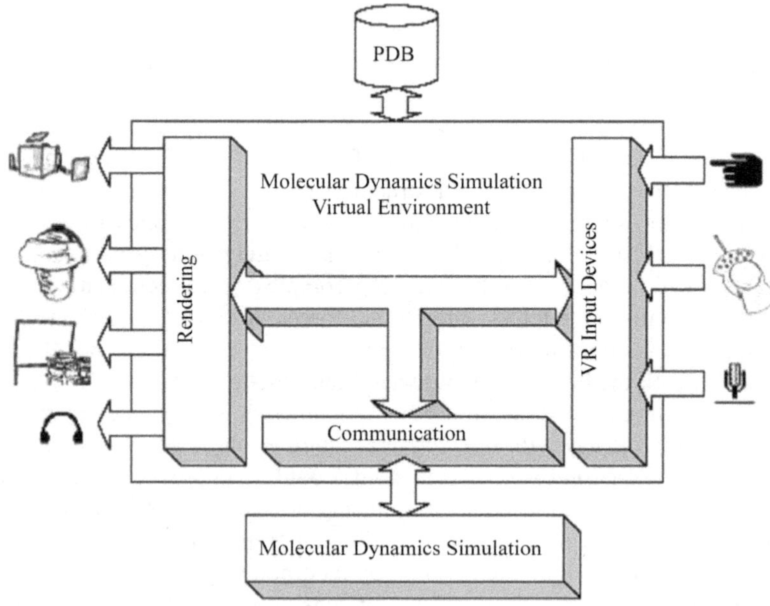

Fig. 1.26 A block diagram showing virtual environment for molecular dynamics simulations (reprinted from: Z. Ai, T. Frohlich)

Computational Steering and Visualization of Complex Molecular Systems

Computational steering is the ability of the user to design or modify a simulation interactively in a virtual environment during run-time, which gives a tremendous advantage over post-simulation visualization and analysis of results. In run-time steering the user does not have to wait till the end of the simulation to see the results of his modifications, instead he can immediately see the result of the interactively changed parameters giving him an opportunity to detect and modify them to steer the simulation to a desired output. In computational steering the user can steer molecular dynamics simulation by applying external forces into the computations. These external forces can help a complex molecular system overcome a potential energy barrier and can even steer the system to a new geometric conformation for further analysis. Thus it provides a great advantage over Targeted Molecular Dynamics [48] which also targets a molecular dynamics simulation to a desired output but in which the user has no control over the simulations once it has started. Another advantage of Steered Molecular Dynamics over conventional molecular dynamics is the possibility of inducing relatively large conformational changes in molecules on nanosecond time scales [49]. Figure 1.27 shows the force induced unfolding of the protein titin [50]. Computational steering of molecular dynamics simulation with the aid of

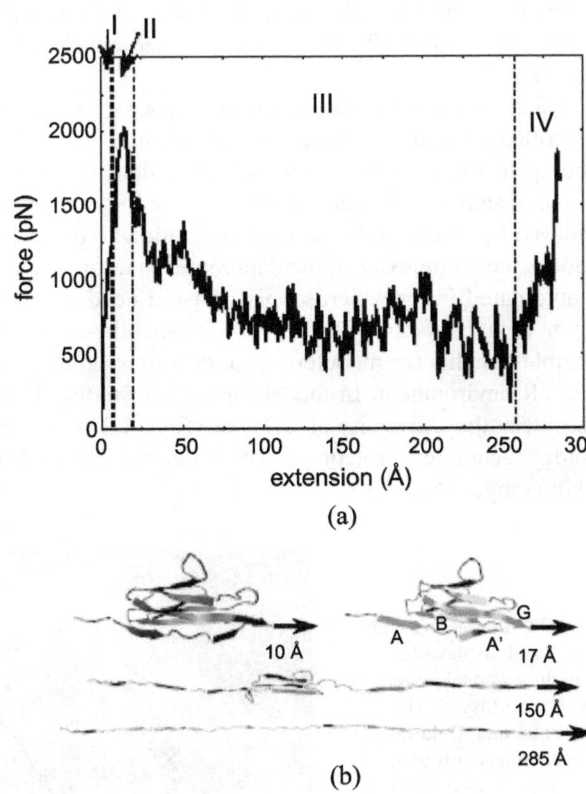

Fig. 1.27 (**a**) Force extension profile from SMD simulations of titin I27 domain with a pulling velocity $v = 0{:}5$ Å/ps. The extension domain is divided into four sections: I. preburst, II. major burst, III. post-burst, IV. pulling of fully extended chain. (**b**) Intermediate stages of the force-induced unfolding. All I27 domains are drawn in the cartoon representation of the folded domain; solvating water molecules are not shown. The four figures at extensions 10 Å, 17 Å, 150 Å, and 285 Å correspond, respectively, to regions I to IV in (**a**) (courtesy of: Dr. Klaus Schulten, Theoretical and Computational Biophysics Group, University of Illinois at Urbana Champaign)

virtual reality visualization can thus help molecular scientists and researchers to explore new models, their structural behaviour and also to study the elastic/mechanical behavior of bio-nanotechnology relevant complex molecular systems.

Current Experimental VR Systems for Bio-nanotechnology

The Virtual Molecular studio at Imperial College London [51] and the Technische Hochschule Darmstadt [52] have done some pioneering research in this field. Also researchers at the Institut für Molekulare Biotechnologie, Germany have set up an internet-based Image Library of Biological Macromolecules [53]. This was one of the first VRML applications in biology. The VRML division of this library contains more than 650 VRML representations of biopolymer structures. One of the challenges in computational structural biology is to enable the efficient use and interoperation of a diverse set of techniques to simulate, analyze, model, and visualize the complex architecture and interactions of macromolecular systems. Scientists at The Scripps Research Institute, California in collaboration with the Central Institute for Applied Mathematics, Germany are developing a 3D graphics package termed SenSitus [54] that is capable of supporting VR devices such as stereo glasses, 3D trackers, and force-feedback (haptic) devices. This development effort will permit scientists to build models, combine atomic and volumetric data, and perform morphing and warping (flexible docking) interactively within a single computational environment.

A force feedback device measures a user's hand position and exerts a precisely controlled force on the hand. The software supports this by calculating forces according to the correlation coefficient of density maps and crystallographic data. The high sampling frequency required for force feedback (refresh rate >1 kHz) is achieved by means of the vector quantization algorithm developed by the group that reduces the complexity of the data representation to manageable levels. In addition to automated fitting, microscopists have a need to evaluate and manipulate docking models interactively "by eye". 3D capabilities and the "physics of touch" offer tangible benefits for modelers who wish to explore a variety of docking situations in a VR environment. In the Visualization Studio of the Delaware Biotechnology Institute at the University of Delaware (Fig. 1.28), research is focused on molecular biology, genomics, proteomics, structural and computational biology, and biomedical imaging.

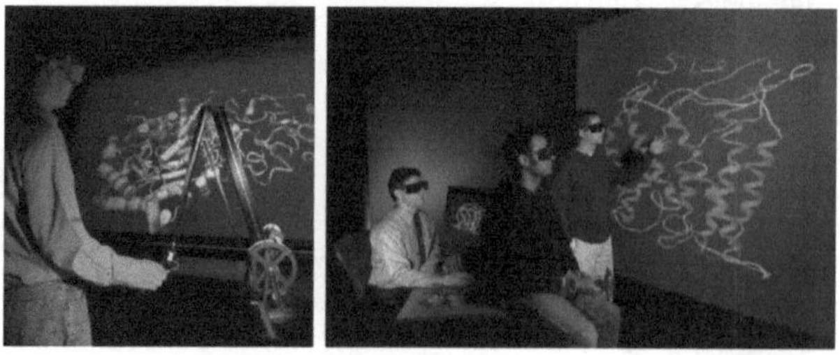

Fig. 1.28 Interacting with a room-sized display of a three-dimensional protein in CAVE (courtesy of: Dr. Karl V. Steiner, Delaware Biotechnology Institute, University of Delaware)

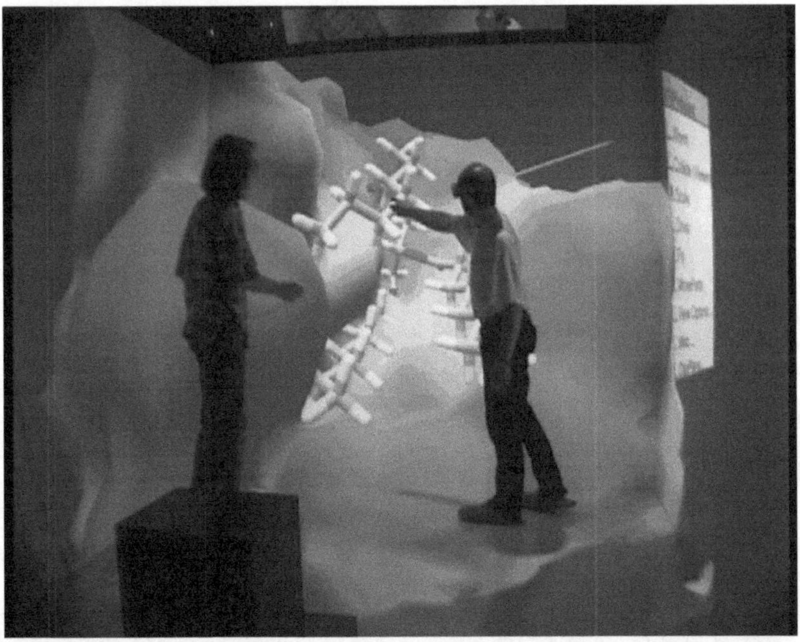

Fig. 1.29 Geometrical representation model for the molecule data (courtesy of: Dr. Juergen Pleiss, HIMM project, University of Stuttgart, Germany)

The Institute's interactive, immersive Visualization Studio consists of a SGI Reality Center system linked to a 15 × 7-foot FakeSpace [10] rear-projection screen. A six-processor Onyx 3200 visualization supercomputer with two graphics pipes drives a pair of Mirage 2000 projectors to deliver an edge- blended image with a total resolution of approximately 2560 × 1024 pixels. The Studio is currently being outfitted with a haptic feedback system to further improve the immersive interface with the visualized environment. The Institute of Technical Biochemistry at the University of Stuttgart (Fig. 1.29) has an ongoing project regarding the application of virtual reality technology in the area of molecular modeling [56]. The Highly Immersive Molecular Modeling (HIMM) [57] project is aimed on the integration of computer aided molecular modeling tools and virtual reality systems like COVISE [58]. Recently, however, another way of visualizing atoms has become available. Researchers in the NAS data analysis group have developed an application called Virtual Mechanosynthesis or VMS [59]. The user of this application sees various collections of atoms floating in the space above the NAS Visualization Laboratory's Immersive Workbench, made by Fakespace Inc. The VMS utility allows the user to see, move, and even "feel" simulated molecular structures in 3-D. A haptic, or force-feedback device can also be used in VMS to interact with the simulation. This device is essentially a mouse mounted on a mechanical arm. It can be pushed around in three dimensions and, by way of a clever arrangement of small motors, it can push back.

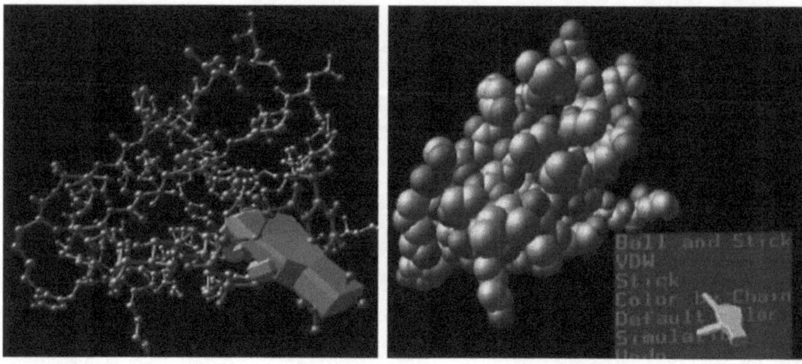

(a) (b)

Fig. 1.30 (**a**) Molecules can be grasped with a cyberglove and moved to a desired position. (**b**) 3-D menu can be activated and selected by a gesture of the cyberglove (reprinted from: Z. Ai, T. Frohlich)

Once users "attach" to an atom using the haptic interface, they can feel the attractive and repulsive forces as this atom pushes and pulls on its neighbors. Another kind of VR interface for molecular dynamics simulation has been developed at the Fraunhofer Institute for Computer Graphics, Germany. This system, called RealMol [60] is implemented to run on CAVE or any other computer system with head mounted display or a stereo projection screen.The MD simulation program for RealMol is NAMD and the communication between the two to exchange molecular data is achieved by RAPP [61] developed at the National Center for Supercomputing Applications, University of Illinois at Urbana-Champaign. The IDEAL Interaction DEvice Abstraction Layer system [62] is used to interface RealMol with the CAVE rendering system. IDEAL is independent of the choice of rendering system, provides an easy to use interface and handles devices for interaction in a virtual environment. With the use of IDEAL a cyberglove can be displayed in the virtual environment as a mapping of hand (Fig. 1.30). Modeling proteins as biological motors as well as designing proteins not found in nature is a rapidly developing field. Keeping this in mind researchers at the University of North Carolina at Chapel Hill have built SMD [63], a system for interactively steering molecular dynamics calculations by adding user-specified external forces into the computation on the fly. Steering implies that the user is able to 'tug' an atom or a group of atoms towards a desired target position. SMD consists of two software components. First one is the software component for performing the molecular dynamics simulations on a molecular system and to study its response to an externally applied steering force.

The display component of SMD is VMD (Fig. 1.31). An example from the University of Washington is given in Fig. 1.32, where the superimposition of multimodal interaction can greatly complement and accelerate the learning process of the operator. For initial demonstration a superoxide dismutase (SOD), an essential enzyme for cellular functioning which exhibits a strong electrostatic funneling effect is modeled. In this scenario the user holds the superoxide radical with the haptic device probe and, as it nears the charge field of SOD, strong forces pull the super-

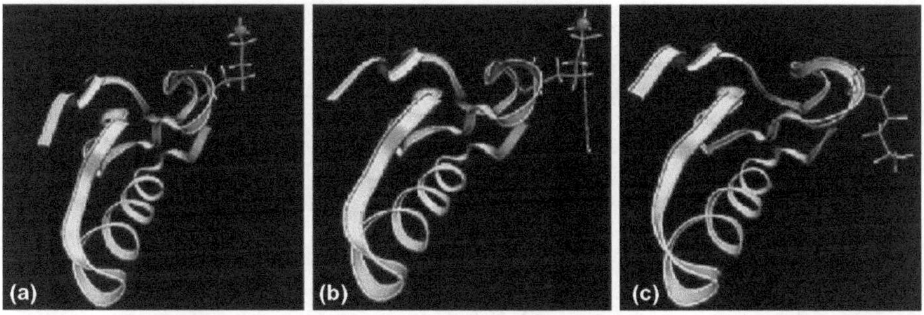

Fig. 1.31 (Color online) (**a**) Specifying a 'tug': selected atom is highlighted as the *red sphere* at upper right. (**b**) Moving the tug: target position follows pointer. (**c**) System response after 300 fs of simulation (courtesy of: Jon Leech, Jan F. Prins, Jan Hermans, University of North Carolina at Chapel Hill)

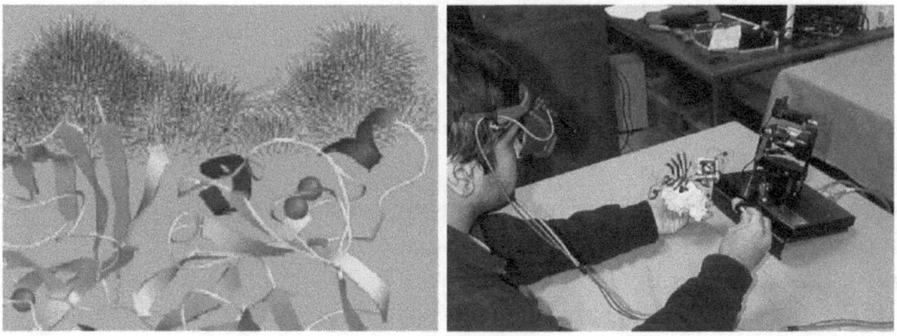

(a) (b)

Fig. 1.32 (**a**) Display of force vector field around active site of superoxide dismutase (SOD) and (**b**) user interacting with SOD models using HMD and PHANToM (courtesy of: Ganesh Shankarnarayan, Human Interface Technology Lab, University of Washington)

oxide free radical toward the Cu and Zn ions at the active site of SOD. At the same time the user sees the secondary structure of the SOD enzyme as an augmented reality overlay on top of the physical model.

They have also developed a Virtual Reality Peripheral Network (VRPN) [64]. The VRPN is a set of classes within a library and a set of servers that are designed to implement a network-transparent interface between application programs and the set of physical devices (tracker, etc.) used in a VR system. The idea is to have a PC or other host at each VR station that controls the peripherals (tracker, button device, haptic device, analog inputs, sound, etc). VRPN provides connections between the application and all of the devices using the appropriate class-of-service for each type of device sharing this link. STALK [65] is a VR based system developed at the University of Illinois at Urbana Champagne and Argonne National Lab for studying the docking of a ligand molecule into a protein binding site.

STALK, shown in Fig. 1.33, uses a parallel genetic algorithm library to search for a low-energy conformation. An interface to the CAVE virtual reality system allows

Fig. 1.33 A user in STALK (courtesy: MCS Division, Argonne National Laboratory)

a scientist to (1) visualize the genetic algorithm's progress and, (2) interact with the algorithm by, for example, changing the position or structure of the drug molecule and restarting the algorithm so that the new parameters are incorporated [66]. The Virtual Biomolecular Environment (VIBE) [67] is a VR based system proposed at Argonne National Lab to provide an environment for Drug Design. It consists of (1) a massively parallel computing to simulate the physical and chemical properties of a molecular system, (2) the Cave Automatic Virtual Environment (CAVE) for immersive display and interaction with the molecular system, and (3) a high-speed network interface to exchange data between the simulation and the CAVE. VIBE enables molecular scientists to have a visual, auditory, and haptic experience with a chemical system, while simultaneously manipulating its physical properties by steering, in real-time, a simulation executed on a supercomputer.

4 Modeling and Characterization Methods

When size goes down to nanoscale or molecular scale, besides the widely known spatial and temporal multiscale problems, new physical properties and coupling effects between mechanical motion and physical, biochemical properties may become dominant, and multiphysics-scale problems arise. Design of bio-nanorobotic systems is a complex and need research on modeling and simulation of physical, chemical, mechanical and biological systems at the nanoscale. These include techniques that mainly cover the range of multidisciplines from atomistic mechanics to continuum mechanics: quantum mechanics, multi-particle simulation, molecular simulation, continuum-based models, stochastic methods, and nanomechanics.

Approaches that make use of more than one such technique and focus on their integration play an important role in this effort.

Current proposed multiscale models for inorganic materials do not integrate the hierarchical design and focus mainly in the multiscale modeling of single components, such as carbon nanotubes [134, 135] or electronic materials [136]. Some multiphysics models have been proposed but the proposed models does not take into consideration all physical interactions, such as quantum mechanics (QM) [137], nanomechanics (NM) [138], or multiphysics parametrization in order to ensure connection between length scales (from atomistic to macroscopic) [139]. Computational methods based on molecular dynamics (MD) are of great interest for investigations of structure, function, and dynamics of structural biology from the atomic level to the macromolecular level [140, 141]. The fundamental issue is to integrate models addressing phenomena from the level of individual gene and cell features through tissue and organ models. Recent approaches are combining multiscale modeling of macromolecular conformational changes to concepts from rigidity and elastic network theory [142, 143]. A good example is given in [144] where the structural dynamics of the lac repressor-DNA complex is described by all-atom molecular dynamics and continuous elastic ribbon. Still, due to their size, biomolecular systems characterized by computational methods like molecular dynamics, are often beyond the reach of real-time feedback for control.

We propose in this thesis work a framework that couple multiscale, multiphysics and virtual reality advanced techniques for the design and modeling of bionanorobotic systems. A three-level modeling paradigm was developed for this purpose. Quantum mechanics (QM) for designed components parametrization, molecular dynamics (MD) for mechanical properties characterization and continuum-model (CM) efficiently integrated with molecular dynamics. As a good illustration of the above multiscale and multiphysics models, a nanorobotic parallel platform composed of biological proteins and inorganic graphite material is designed, modeled and optimized. Simulation results show clearly the main advantages of the proposed method in terms of computational time, accuracy of simulation and energy optimization at the pre-project stage. Furthermore, co-prototyping methods centered on modeling algorithms and experimental measurement data are developed in order to optimize and prototype multiwalled carbon nanotubes-based nanodevices at the post-project stage.

5 Conclusion

The technological progress allow to observe, modify and to control the behavior of the natural nanomachine and to use it for new applications and tasks such as nanomedicine, mechanical engineering at the nanometer scale. Also it became possible to design new nanodevices for quite specific applications. The design of new nanodevices require the characterization, prediction of device behavior and prototyping. Thus an analysis and study allow the classification and to choose the suitable nanomaterials candidates to fabricate a nanorobot. In this thesis work we propose novel methods of design dedicated to the nanorobotic and bio-nanorobotic devices. For the design of the bio-nanorobotic structures our approach consists in coupling a multiscale modeling with virtual reality techniques, this method allow to handle the molecular elements in an interactive 3D environment and to characterize these elements through realistic models. For the design of the nanorobotic structures (mainly based on carbon nanotubes), we propose a method allowing to couple realistic models with an experimental platform, termed throughout the manuscript co-prototyping method. The organization of thesis chapter's is as following:

- First, Chap. 2 presents the different methodologies that have been developed in order to design and to characterize bionano and nanorobotic components, structures and systems.
- Then, Chap. 3 proposes novel design, modeling, and computational studies of biological components for dedicated to bio-nanorobotics structures.
- Finally, Chap. 4 proposes different co-prototyping methods for optimization of multiwalled carbon-based nanodevices realized in collaboration with the Institute of Robotics and Intelligent Systems, ETH Zurich.

References

1. Calvacanti, A., Shririnzadeh, B., Freitas, R., Hogg, T.: Nanorobot architecture for medical target identification. Nanotechnology **19**, 1–15 (2008)
2. Requicha, A.: Instrumented cellular systems. Nanomed. Nanotechnol. Biol. Med. **2**(4), 278 (2006)
3. Hogg, T., Kuekes, P.J.: Mobile microscopic sensors for high resolution in vivo diagnostics. Nanomed. Nanotechnol. Biol. Med. **2**(4), 239–247 (2006)
4. Dong, L.X., Nelson, B.J., Fukuda, T., Arai, F.: Towards linear nano servomotors. IEEE Trans. Autom. Sci. Eng. **3**(3), 228–235 (2006)
5. Subramanian, A., Dong, L.X., Nelson, B.J.: Batch fabrication of carbon nanotube bearings. Nanotechnology **18**, 075703 (2007)
6. Chirikjian, G.S., Kazerounian, K., Mavroidis, C.: Analysis and design of protein based nanodevices: challenges and opportunities in mechanical design. J. Mech. Des. **127**, 695–698 (2005)
7. Yesin, K.B., Vollmers, K., Nelson, B.J.: Modeling and control of unthered biomicrorobots in a fluidic environment using electromagnetic fields. Int. J. Rob. Res. **25**(5–6), 527–536 (2006)
8. Mathieu, J.-B., Beaudoin, G., Martel, S.: Method of propulsion of a ferromagnetic core in the cardiovascular system through magnetic gradients generated by an MRI system. IEEE Trans. Biomed. Eng. **53**(2), 292–299 (2006)
9. Behkam, B., Sitti, M.: Design methodology for biomimetic propulsion of miniature swimming robots. J. Dyn. Syst. Meas. Control **128**, 36–43 (2006)
10. Martel, S., Tremblay, C., Ngakeng, S., Langlois, G.: Controlled manipulation and actuation of microobjects with magnetotactic bacteria. Appl. Phys. Lett. **89**, 233804 (2006)
11. Hamdi, M., Ferreira, A.: DNA nanorobotics. Microelectron. J. (2008)
12. Steager, E., Kim, C.B., Patel, J., Bith, S., Naik, C., Reber, L., Kim, M.J.: Control of microfabricated structures powered by flagellated bacteria using phototaxis. Appl. Phys. Lett. **90**, 263901 (2006)
13. Shingyoji, C., Higuchi, H., Yoshimura, M., Katayama, E., Yanagida, T.: Dynein arms are oscillating force generators. Nature **393**, 711–714 (2001)

14. Mitsuya, Y.: Significance of micro-nanomechatronics for an information-based society. In: The Fourth Symposium, Micro-Nanomechatronics for Information-Based Society, pp. 29–31, 31 Oct.–3 Nov. 2004
15. Iijima, S.: Nature (Lond.) **354**(6348), 56–58 (1991)
16. Hamada, N., Sawada, S.I., Oshiyama, A.: New onedimensional conductors, graphite microtubules. Phys. Rev. Lett. **68**, 1579–1581 (1992)
17. Dresselhaus, M.S., Dresselhaus, G., Eklund, P.C.: Science of Fullerenes and Carbon Nanotubes. Academic Press, San Diego (1995)
18. Tersoff, J., Ruoff, R.S.: Structural properties of a carbon-nanotube crystal. Phys. Rev. Lett. **73**, 676–679 (1994)
19. Wang, N., Tang, Z.K., Li, G.D., Chen, J.S.: Singlewalled 4 Å carbon nanotube arrays. Nature **408**, 50–51 (2000)
20. Haddon, R.C.: Chemistry of the fullerenes: the manifestation of strain in a class of continuous aromatic molecules. Science **261**, 1545–1550 (1993)
21. Hamdi, M., Sharma, G., Ferreira, A., Mavroidis, D.: Characterization of protein based spring-like elastic joints for biorobotic applications. In: IEEE International Conference on Robotics and Automation, pp. 1794–1799, Orlando, FL, May 15–19, 2006
22. Hamdi, M., Sharma, G., Ferreira, A., Mavroidis, D.: Prototyping bio-nanorobots using molecular dynamics simulation and virtual reality. Microelectron. J. **39**, 190–201 (2008)
23. Mahadevan, L., Matsudaira, P.: Mobility powered by supramolecular springs and ratchets. Science **288**, 95–99 (2000)
24. Sakai, L.Y., Keene, D.R., Engvall, E.: Fibrillin, a new 350-kD glycoprotein, is a component of extracellular microfibrils. J. Cell Biol. **103**, 2499–2509 (1986)
25. Treasy, M.M.J., Ebbesen, T.W., Gibson, J.M.: Nature (Lond.) **381**, 678 (1996)
26. Wong, E.W., Sheehan, P.E., Lieber, C.M.: Science (Wash.) **277**, 1971 (1997)
27. Zheng, Q., Liu, J.Z., Jiang, Q.: Phys. Rev. B Condens. Matter **65**, 245409 (2002)
28. Tu, Z.C., Hu, X.: Phys. Rev. B Condens. Matter **72**, 033404 (2005)
29. Maslov, L.: Nanotechnology **17**, 2475 (2006)
30. Saito, R., Matsuo, R., Kimura, T., Dresselhaus, G., Dresselhaus, M.S.: Chem. Phys. Lett. **348**, 187 (2001)
31. Lozovik, Y.E., Minogin, A.V., Popov, A.M.: Phys. Lett. A **313**, 112 (2003)
32. Lozovik, Y.E., Minogin, A.V., Popov, A.M.: Pisma Z. Éksp. Teor. Fiz. **77**(11), 759 (2003). [JETP Lett. **77**(11), 631 (2003)]
33. Lozovik, Y.E., Popov, A.M.: Fullerenes, nanotubes. Carbon Nanostruct. **12**, 485 (2004)
34. Fennimore, A.M., Yuzvinsky, T.D., Han, W.Q., Fuhrer, M.S., Cumings, J., Zettl, A.: Nature (Lond.) **424**, 408 (2003)
35. Bourlon, B., Glatti, D.C., Forro, L., Bachfold, A.: Nano Lett. **4**, 709 (2004)
36. Petsko, G.A., Ringe, D.: Protein Structure and Function: From Sequence to Consequence. New Science Press, London (2004)
37. Noji, H., et al.: Direct observation of the rotation of F1-ATPase. Nature **386**(6622), 299–302 (1997)
38. Lubert, S.: Biochemistry, 4th edn. Freeman, New York (1995)
39. Noji, H.: The rotary enzyme of the cell: the rotation of F1-ATPase. Science **282**, 1844–1845 (1998)
40. Vale, R.D., Milligan, R.A.: The way things move: looking under the hood of molecular motor proteins. Science **288**, 88–95 (2000)
41. Pallen, M.J., Matzke, N.J.: From the origin of species to the origin of bacterial flagella. Nat. Rev. Microbiol. **4**, 784–790 (2006)
42. Mavroidis, C., Dubey, A., Yarmush, M.: Molecular machines. Annu. Rev. Biomed. Eng. **6**, 363–395 (2004)
43. Berman, H.M., Westbrook, J., Feng, Z., Gilliland, G., Bhat, T.N., Weissig, H., Shindyalov, I.N., Bourne, P.E.: The protein data bank. Nucleic Acids Res. **28**, 235–242 (2000)
44. Sühnel, J.: Virtual reality modeling for structural biology. Institut fur Molekulare Biotechnologie, Postfach 100813, D-07708 Jena/Germany

45. Sankaranarayanan, G., Weghorst, S., Sanner, M., Gillet, A., Olson, A.: Role of haptics in teaching structural molecular biology. In: 1th Symposium on Haptic Interfaces for Virtual Environment and Teleoperator Systems, p. 365, Los Angeles, California, March 22–23, 2003
46. Schlitter, J.: Targeted molecular dynamics: a new approach for searching pathways of conformational transitions. J. Mol. Graph. **12**, 84–89 (1994)
47. Izrailev, S., Stepaniants, S., Isralewitz, B., Kosztin, D., Lu, H., Molnar, F., Wriggers, W., Schulten, K.: Targeted molecular dynamics: a new approach for searching pathways of conformational transitions. J. Mol. Graph. **12**, 84–89 (1994)
48. Izrailev, S., Stepaniants, S., Isralewitz, B., Kosztin, D., Lu, H., Molnar, F., Wriggers, W., Schulten, K.: Steered molecular dynamics. In: Computational Molecular Dynamics: Challenges, Methods, Ideas, pp. 39–65. Springer, Berlin (1998)
49. Casher, O., Leach, C., Page, C.S., Rzepa, H.S.: Advanced VRML based chemistry applications: a 3D molecular hyperglossary. J. Mol. Struct. **368**, 49 (1996)
50. Casher, O., Leach, C., Page, C.S., Rzepa, H.S.: Virtual reality modelling language (VRML) in chemistry. Chem. Br. **34**, 26 (1998)
51. Casher, O., Rzepa, H.S.: A chemical collaboratory using explorer eyechem and the common client interface. Comput. Graph. **29**, 52 (1995)
52. Wriggers, W., Birmanns, S.: Using situs for flexible and rigid-body fitting of multi-resolution single molecule data. J. Struct. Biol. **133**, 193–202 (2001)
53. ImmersaDesk. http://www.fakespace.com/workdesk1.shtml, 2003 Fakespace Systems Inc.
54. Haase, H., Strassner, J., Dai, F.: VR techniques for the investigation of molecule data. Comput. Graph. **20**(2), 207–217 (1996). Special Issue on Virtual Reality
55. Drees, R.C., Pleiss, J., Roller, D., Schmid, R.D.: Highly immersive molecular modeling (HIMM): an architecture for the integration of molecular modeling and virtual reality. In: German Conference on Bioinformatics, pp. 190–192, Sep.–Oct. 1996
56. Collaborative Visualization and Simulation Environment (COVISE): http://www.hlrs.de/organization/vis/covise/. Last updated Mar. 30, 2001
57. Levit, C., Bryson, S.T., Henze, C.E.: Virtual mechanosynthesis. In: Fifth Foresight Conference on Molecular Nanotechnology, California, Nov. 1997
58. Stone, J.E., Gullingsrud, J., Schulten, K., Grayson, P.: A system for interactive molecular dynamics simulation. In: 2001 ACM Symposium on Interactive 3D Graphics, pp. 191–194, New York, USA (2001)
59. Humphrey, W.F., Dalke, A., Schulten, K.: VMD—visual molecular dynamics. J. Mol. Graph. **14**, 33–38 (1996)
60. Roth, M., Frohlich, T.: IDEAL Interaction DEvice Interaction Layer User's Manual. Fraunhofer-IGD, internal report (1997)
61. IDEAL Interaction DEvice Interaction Layer User's Manual. Fraunhofer-IGD, internal report (1997)
62. Leech, J., Prins, J.F., Hermans, J.: SMD: visual steering of molecular dynamics for protein design. IEEE Comput. Sci. Eng. **4**, 38–45 (1996)
63. Taylor, R.M. II, Hudson, T.C., Seeger, A., Weber, H., Juliano, J., Helser, A.T.: VRPN: a device-independent, network-transparent VR peripheral system. In: ACM VRST (2001)
64. Levine, D., Facello, M., Hallstrom, P., Reeder, G., Walenz, B., Stevens, F.: STALK: an interactive virtual molecular docking system. IEEE Comput. Sci. Eng. (1996)
65. Cruz-Neira, C., Langley, R., Bash, P.A.: VIBE: a virtual biomolecular environment for interactive molecular modeling. Comput. Chem. **20**(4), 469 (1996)
66. Arthur, K., Preston, T., Taylor, R.M. II, Brooks, F.P., Jr., Whitton, M.C., Wright, W.V.: Designing and building the PIT: a head-tracked stereo workspace for two users. In: 2nd International Immersive Projection Technology Workshop, Ames, Iowa, 11–12 May 1998
67. Prins, J.F., Hermans, J., Mann, G., Nyland, L.S., Simons, M.: A virtual environment for steered molecular dynamics. Future Gener. Comput. Syst. **15**(4), 485–495 (1999)
68. Saenger, W.: Principles of Nucleic Acid Structure. Springer, New York (1984)
69. Watson, J.D., et al.: Molecular Biology of the Gene, 4th edn. Benjamin/Cummings, Redwood City (1987)

70. Nishinaka, T., Ito, Y., Yokoyama, S., Shibata, T.: Proc. Natl. Acad. Sci. USA **94**, 6623 (1997)
71. Nishinaka, T., Shinohara, A., Ito, Y., Yokoyama, S., Shibata, T.: Proc. Natl. Acad. Sci. USA **95**, 11071 (1998)
72. Lt'eger, J.F., Robert, J., Bourdieu, L., Chatenay, D., Marko, J.F.: Proc. Natl. Acad. Sci. USA **95**, 12295 (1998)
73. Stewart, L., Redinbo, M.R., Qiu, X., Hol, W.G.J., Champoux, J.J.: Science **279**, 1534 (1998)
74. Rybenkov, V.V., Ullsperger, C., Vologodskii, A.V., Cozzarelli, N.R.: Science **277**, 690 (1997)
75. Yan, J., Magnasco, M.O., Marko, J.F.: Nature (Lond.) **401**, 932 (1999)
76. Smith, S.B., Finzi, L., Bustamante, C.: Science **258**, 1122 (1992)
77. Bensimon, D., Simon, A.J., Croquette, V., Bensimon, A.: Phys. Rev. Lett. **74**, 4754 (1995)
78. Cluzel, P., Lebrun, A., Heller, C., Lavery, R., Viovy, J.-L., Chatenay, D., Caron, F.: Science **271**, 792 (1996)
79. Smith, S.B., Cui, Y., Bustamante, C.: Science **271**, 795 (1996)
80. Strick, T.R., Alleman, J.-F., Bensimon, D., Bensimon, A., Croquette, V.: Science **271**, 1835 (1996)
81. Strick, T.R., Croquette, V., Bensimon, D.: Proc. Natl. Acad. Sci. USA **95**, 10579 (1998)
82. Allemand, J.F., Bensimon, D., Lavery, R., Croquette, V.: Proc. Natl. Acad. Sci. USA **95**, 14152 (1998)
83. Lt'eger, J.F., Romano, G., Sarkar, A., Robert, J., Bourdieu, L., Chatenay, D., Marko, J.F.: Phys. Rev. Lett. **83**, 1066 (1999)
84. Bustamante, C., Marko, J.F., Siggia, E.D., Smith, S.: Science **265**, 1599 (1994)
85. Marko, J.F., Siggia, E.D.: Macromolecules **28**, 8759 (1995)
86. Yu, M.F., Lourie, O., Dyer, M.J., Moloni, K., Ruoff, R.S.: Science **287**, 637 (2000)
87. Yu, M.F., Yakobson, B.I., Ruoff, R.S.: J. Phys. Chem. B **104**, 8764 (2000)
88. Cumings, J., Zettl, A.: Science **289**, 602 (2000)
89. Tuzun, R.E., Noid, D.W., Sumpter, B.G.: Nanotechnology **6**, 52 (1995)
90. Srivastava, D.W.: Nanotechnology **8**, 186 (1997)
91. Forro, L.: Science **289**(5479), 560 (2000)
92. Zheng, Q., Jiang, Q.: Phys. Rev. Lett. **88**, 045503 (2002)
93. Kresse, H.P., et al.: Four-helix bundle topology re-engineered: monomeric Rop protein variants with different loop arrangements. Protein Eng. **14**(11), 897–901 (2001)
94. Balzani, V., et al.: Molecular Devices and Machines. Wiley–VCH, New York–Weinheim (2003)
95. Finer, J.T., et al.: Nature **368**, 113 (1994)
96. Svoboda, K., et al.: Nature **365**, 721 (1993)
97. Howard, J.: Mechanics of Motor Proteins and the Cytoskeleton. Sinauer, Sunderland (2001)
98. Wuite, G.J.L., et al.: Nature **404**, 103 (2000)
99. Beissenhirtz, M.K., Willner, I.: Org. Biomol. Chem. **4**, 3392 (2006)
100. Mandal, M., Breaker, R.R.: Nat. Rev. Mol. Cell Biol. **5**, 451 (2004)
101. Buranachai, C., et al.: Nano Lett. **6**, 496 (2006)
102. Müller, B.K., et al.: Nano Lett. **6**, 2814 (2006)
103. Liu, D.S., et al.: J. Am. Chem. Soc. **128**, 2067 (2006)
104. Liedl, T., et al.: Angew. Chem. Int. Ed. **45**, 5007 (2006)
105. Hazarika, P., et al.: Angew. Chem. Int. Ed. **43**, 6469 (2004)
106. Yan, H., et al.: Nature **415**, 62 (2002)
107. Feng, L.P., et al.: Angew. Chem. Int. Ed. **42**, 4342 (2003)
108. Shu, W.M., et al.: J. Am. Chem. Soc. **127**, 17054 (2005)
109. Yang, X.P., et al.: Biopolymers **45**, 69 (1998)
110. Yurke, B., et al.: Nature **406**, 605 (2000)
111. Simmel, F.C., Yurke, B.: Phys. Rev. E **63**, 041913 (2001)
112. Mitchell, J.C., Yurke, B.: DNA Computing, 7th International Workshop on DNA-Based Computers. LNCS, vol. 2340, p. 258. Springer, Heidelberg (2002)
113. Shen, W.Q., et al.: Angew. Chem. Int. Ed. **43**, 4750 (2004)

114. Chen, Y., et al.: Angew. Chem. Int. Ed. **43**, 3554 (2004)
115. Chhabra, R., et al.: Nano Lett. **6**, 978 (2006)
116. Seeman, N.C.: Trends Biochem. Sci. **30**, 119 (2005)
117. Shin, J.S., Pierce, N.A.: J. Am. Chem. Soc. **126**, 10834 (2004)
118. Sherman, W.B., Seeman, N.C.: Nano Lett. **4**, 1203 (2004)
119. Tian, Y., Mao, C.: J. Am. Chem. Soc. **126**, 11410 (2004)
120. Yin, P., et al.: Angew. Chem. Int. Ed. **43**, 4906 (2004)
121. Tian, Y., et al.: Angew. Chem. Int. Ed. **44**, 4355 (2005)
122. Pei, R., et al.: J. Am. Chem. Soc. **128**, 12693 (2006)
123. Li, J.W.J., Tan, W.H.: Nano Lett. **2**, 315 (2002)
124. Makita, N., et al.: Nucleic Acids Symp. Ser. **48**, 173 (2004)
125. Fahlman, R.P., et al.: Nano Lett. **3**, 1073 (2003)
126. Anelli, P.-L., Spencer, N., Stoddart, J.F.: A molecular shuttle. J. Am. Chem. Soc. **113**, 5131–5133 (1991)
127. Balzani, V.V., Credi, A., Raymo, F.M., Stoddart, J.F.: Artificial molecular machines. Angew. Chem. Int. Ed. Engl. **39**, 3348–3391 (2000)
128. Schalley, C.A., Beizai, K., Vogtle, F.: On the way to rotaxane-based molecular motors: studies in molecular mobility and topological chirality. Acc. Chem. Res. **34**, 465–476 (2001)
129. Gatti, F.G., Leon, S., Wong, J.K.Y., Bottari, G., Altieri, A., et al.: Photoisomerization of a rotaxane hydrogen bonding template: light-induced acceleration of a large amplitude rotational motion. Proc. Natl. Acad. Sci. USA **100**, 10–14 (2003)
130. Bermudez, V.V., Capron, N., Gase, T., Gatti, F.G., Kajzar, F., et al.: Influencing intramolecular motion with an alternating electric field. Nature **406**, 608–611 (2000)
131. Fyfe, M.C.T., Stoddart, J.F.: Synthetic supramolecular chemistry. Acc. Chem. Res. **30**, 393–401 (1997)
132. Ashton, P.R., Goodnow, T.T., Kaifer, A.W., Reddington, M.V., Slawin, A.M.Z., Spencer, N., Stoddart, J.F., Vicent, C., Williams, D.J.: A [2]catenane made to order. Angew. Chem. Int. Ed. Eng. **28**, 1396–1399 (1989)
133. Deleuze, M.S.: Can benzylic amide [2]catenane rings rotate on graphite? J. Am. Chem. Soc. **122**, 1130–1143 (2000)
134. Maiti, A.: Multiscale modeling with carbon nanotubes. Microelectron. J. (2006)
135. Liu, Y.J., Chen, X.L.: Continuum models of carbon nanotube-based composites using the boundary element method. J. Bound. Elem. **1**(2), 316–335 (2003)
136. Karakasidis, T.E., Charitidis, C.A.: Multiscale modeling in nanomaterials science. Mater. Sci. Eng. (2006)
137. Frenkel, D., Smit, B.: Understanding Molecular Simulation: From Algorithms to Applications, 2nd edn. Academic Press, San Diego (2001)
138. Ghoniem, N.M., Busso, E.P., Kioussis, N., Huang, H.: Multiscale modelling of nanomechanics and micromechanics: an overview. Philos. Mag. **83**(31–34), 3475–3528 (2003)
139. Braatz, R.D., et al.: Perspectives on the design and control of multiscale systems. J. Process Control **16**, 193–204 (2006)
140. Feller, E., Zhang, Y.H., Pastor, R.W., Brooks, B.R.: Constant pressure molecular dynamics simulationthe Langevin piston method. Chem. Phys. J. **103**, 4613–4621 (1995)
141. Hetherington, J., et al.: Addressing the challenges of multiscale model management in systems biology. Comput. Chem. Eng. (2006)
142. Ahmed, A., Gohlke, H.: Multiscale modeling of macromolecular conformational changes combining concepts from rigidity and elastic network theory. Proteins Struct. Funct. Bioinf. **63**, 1038–1051 (2006)
143. Feig, M., Karanicolas, J., Brooks, C.L.: MMTSB Tool Set: enhanced sampling and multiscale modeling methods for applications in structural biology. J. Mol. Graph. Model. **22**, 377–395 (2004)
144. Villa, E., Balaeff, A., Schulten, K.: Structural dynamics of the lac repressor-DNA complex revealed by a multiscale simulation. Proc. Natl. Acad. Sci. USA **102**(19), 6783–6788 (2005)

Chapter 2
Methodology of Design and Characterization of Bionano- and Nanorobotic Devices

1 Introduction

Nanorobotics is an emerging interdisciplinary technology area raising new scientific challenges and promising revolutionary advancements in applications such as medicine, biology and industrial manufacturing. Nanorobots could be defined as intelligent systems with overall dimensions at or below the micrometer range that are made of assemblies of nanoscale passive or active components with individual dimensions ranging between 1 to 100 nm. Nanorobots would be able to perform at least one of the following actions: actuation, sensing, signaling, information, processing, intelligence, swarm behavior at the nano scale. Bio-nanorobots are nanorobots using various biological elements—whose function at the cellular and sub-cellular level creates a motion, force or a signal—as machine components that perform the same function in response to a biological stimuli.

Since the development of these molecular machines is a relatively new research field, the derivation of computational to model and analyze their performance is still at an early stage. To design high-performance nanorobotic structures and systems, fundamental, applied, and experimental research must be performed. One of the most challenging problems in bionano- and nanorobotic design is the topology-architecture-configuration synthesis, system integration and optimization. Towards this end, the selection of analytical and numerical methods, computation algorithms, model accuracy and simulation environment is crucial for virtual prototyping and optimization at an early-stage of development. Indeed, modeling allows the designer to devise novel phenomena and new operational principles guaranteeing synthesis of nanorobotic components, subsystems and systems with enhanced functionalities. The computational studies aim at developing models that can be used in understanding the mechanism of the proposed robotic nanostructure, in predicting its feasibility and performance in terms of energy, force, speed, and other kinematic and dynamic criteria as well as perform optimal design procedures to identify the optimal structure performances. Many novel concepts and designs have been first proposed based on modeling and simulations, and then were followed by their realization or verification through experiments.

M. Hamdi, A. Ferreira, *Design, Modeling and Characterization of Bio-Nanorobotic Systems*,
DOI 10.1007/978-90-481-3180-8_2,

In this chapter we investigated computational nanotechnology methodologies, i.e., physics-, biology- and chemistry-based modeling and simulation, dedicated to the design and optimization of bio-nano structures. An attempt has been made to emphasize the paradigm of how computational nanotechnology can be used to not only understand and characterize the systems under experimental investigation (termed, *co-prototyping design*), but also as to how predictions can be made for novel applications (termed, *interactive multiscale design*), because there is a strong coupling between the structural, mechanical, chemical and biological properties at multiple length–time-scale. Two prototyping methodologies are proposed and discussed:

1. *Interactive multiscale design*: The proposed multiscale methodology hierarchical simulation techniques considers virtual reality techniques coupled to computational quantum mechanics, molecular dynamics and continuum models. The interactive multiscale simulation methodology allows to simulate, analyze, model, and visualizes the complex architecture and interactions of biological nanorobotic components. The goal is to prototype virtually bio-nanorobotic systems and control their movements in their biological environment before experimentation.
2. *Co-prototyping design*: The proposed co-prototyping methodology is an approach to the development of nanostructures composed of experimental characterization and computational nanotechnology modeling (ab-initio and molecular dynamics simulation methods) towards optimized, reliable and reusable nanorobotic structures.

This chapter begins with a section (Sect. 2) describing the technical and simulation aspects of the interactive virtual multiscale simulation system. First, we consider the hardware configuration and virtual reality input devices, and then the multiscale modeling methods and prototyping methodology. Then, a section (Sect. 3) presents the co-prototyping methodology and simulation aspects, including requirements on the use of semi- and empirical modeling methods. Concluding remarks are given in Sect. 4.

2 Design and Characterization Methodology of Biological Nanodevices

Objective

Recent advances in understanding how biomolecular motors work has raised the possibility that they might find applications as protein-based nanomachines. For example, they could be used as molecule-sized robots that are able to apply forces and manipulate objects in the nanoscale world. Thus biomolecular motors could form the basis of bottom-up approaches for constructing active structuring and maintenance at the nanometer scale. Protein is the most versatile of the natural bionanomachines. As an example, the most familiar motor is the protein myosin [1] or dynein [2] which moves along filaments, formed through the protein actin, to drive the contraction of muscles. In addition, there are compliance devices such as spring-like proteins called fibronectin [3] and vorticellids [4], as well as synthetic contractile plant polymers [5] which can act as compliant joints in molecule-size robotic systems. The idea is to use biomolecular motors as the actuators of such bionanorobots, where the structural elements are carbon nanotubes (CNTs), while the joints are formed by appropriately designed biological spring elements [6]. To achieve these long-term goals, prototyping tools based on molecular dynamics (MD) simulators should be developed in order to understand the molecular mechanics of proteins and develop dynamic and kinematic models to study their performances and control aspects. The ability to visualize the atom-to-atom interaction in real-time and see the results in a fully immersive 3D environment is an additional feature of such simulations [7]. Virtual reality (VR) technology is applied here, which not only provides immersive visualization but also gives an added functionality of CAD-based design, simulation, navigation and interactive manipulation of molecular graphical objects. Using virtual biological environments, the operator can design and characterize through physical simulation the behavior of bio-nanorobots. Adding haptic interaction, the operator can explore and prevent the problems of molecular robots in their native environment. Based on VR technology and MD simulators, our long-term goal is to prototype virtually bio-nanorobotic systems and control their movements in their biological environment.

Overall CAD-design Methodology

Considering the above mentioned research issues, we decided to adopt VR technologies coupled to computational molecular dynamics and quantum molecular dynamics to simulate, analyze, model, and visualizes the complex architecture and interactions of bio-nanorobotic components. Real-time exploration is what attracts researchers to develop man-machine interfaces for nanoscale manipulation that use haptic display technologies. The developed simulation system presented in Fig. 2.2 permits manipulation, connection and assembly of bio-nanorobotic components in molecular dynamics simulation with real-time VR devices such as stereo glasses, 3D trackers, force-feedback devices and 3D graphical display.

Nanoassembly methodology requires to address the assembly process systematically to design it optimally. Two major assembly processes are currently employed at the nanoscale. Robotic nanoassembly, much like its macroscopic counterpart, involves positioning and joining of components. Nanocomponents are positioned by using the atomic force microscope (AFM) as a robot, essentially by pushing the components mechanically and joining them chemically, by using linkers such as DNA complementary strands, or by depositing additional material over the templates defined by the nanocomponents [8, 9]. Mass production is best approached by programmed self-assembly, using scaffolds designed by algorithmic methods. Several self-assembly approaches have been proposed based on various forces such

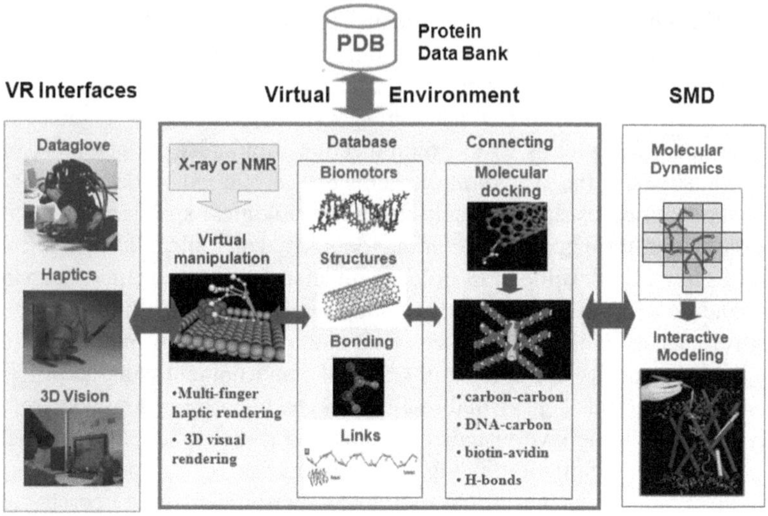

Fig. 2.1 Design methodology of bio-nanostructures

as capillary force, hydrophobic force, electrostatic force and so on [10, 11]. The controlled nanoassembly approach is based on hierarchical theoretical and computational assembly methods depicted on the flowchart shown in Fig. 2.2:

1. To connect and assemble interactively the molecular components of the bionanorobotic structure through a data glove which can be displayed in the virtual environment as a mapping of hand. It is composed of fully enclosed fiber optic bend sensors that provides up to 18 high-accuracy joint-angle measurements allowing to capture all assembly hand postures.
2. To virtually steer the protein molecules using a haptic device. The haptic device is virtually bound to an atom of a conformationally stable subset of the molecule. This system would allow applying stretching, shearing and bending mechanical constraints on the protein.
3. To distinguish possible conformational paths, feel the energy barriers associated with each path and choose the optimal path/strategy.
4. To control the interaction mechanism in a solvent. Protein folding and unfolding can be controlled by assessing the influence of pH value, temperature, electron density or salt concentration.

The software architecture is composed of three software modules: (i) a haptic device module controlled by a computer that generates the force environment, (ii) an interactive molecular environment (IME) module for manipulation, attachment and assembly of nanocomponents in 3-D virtual world and (iii) a molecular dynamics (MD) module for determining the effects of force application.

As shown in Fig. 2.2, steering forces are issued from the haptic device which measures a user's hand position and exerts a precisely controlled force on the hand in order to apply different mechanical constraints, force and energy fields on the

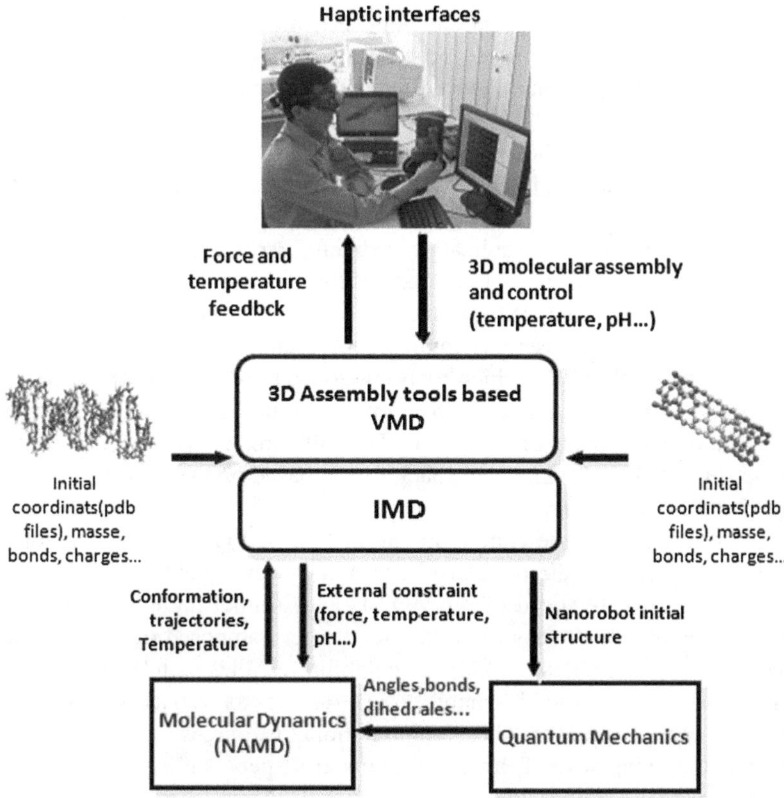

Fig. 2.2 Basic concept of virtual environment and haptics technology coupled to multiphysics computational methods for bio-nanorobotic design

virtual model in order to prototype bionanorobotic components. The haptics module sends command data (position, velocity, force) and receives force feedback signals from the IME module.

The main role of the interactive molecular environment module (IME) module is to gather the nanorobotic components (proteins, DNA, CNT, graphite, ...) in a one 3-D representative space called visual molecular dynamics (VMD) [12]. VMD software is a molecular visualization program for displaying, animating and analyzing large bimolecular systems using 3-D graphics and built-in scripting. The VRPN software package [13] was used to connect VMD to the PHANToM Desktop device. The Data Glove input device is also supported by VRPN. The user manipulates in an intuitive way the attachment and assembly of the different molecular components within VMD environment. As the linkers and/or connectors between multi-domain proteins are not always clear during complex assemblies, *Pdbswiss* software allows the analysis of structural protein and/or nanostructure misalignments (rotation, translation, orientation). When the 3-D structures are aligned, the molecular attachment between the different components is achieved via *VegaZZ* software.[1] Finally, the a molecular structure will be generated, associated to the CAD-designed bio-nano robotic structure: initial coordinates, force field parameters, structure solvation and ionization.

These data are sent to the molecular dynamics (MD) module through a communication protocol named interactive molecular dynamics (IMD) [14]. The IMD interface consists of a small set of C-callable functions which can be adapted to any molecular dynamics and/or visualization program. Periodically, VMD checks for a coordinate set from the MD program, updates the representation geometry if a new coordinate set was received, redraws the screen, and updates the restraint position of the haptic device. If the user has applied a force through the haptic device, VMD routes this force to NAMD2, which then integrates the force into the equations of motion for the molecule. Updates to the haptic restraint point are made when VMD receives a new coordinate set from NAMD2 software; while awaiting an update, the haptic server applies smooth feedback forces based on the most recent restraint point position.

The main program of the computational is NAMD2 [15], a fast, scalable, program which implements the popular CHARMM27 force field [16] for molecular dynamics. The absolute speed of NAMD2 is an essential ingredient of a responsive interactive system. It runs on a parallel cluster composed of 10 Pentium IV–4.1 GHz personal computers. NAMD's scalability enables us to use the computational module to study large bio-nanorobotic structures as well as small ones, given enough computational power. These computational resources allow the simulation of systems containing several thousand atoms at a speed sufficient for interaction ($\sim$400 Hz).

When size goes down to nanoscale or molecular scale, besides the widely known spatial and temporal multiscale problems, new physical properties and coupling effects between mechanical motion and physical, biochemical properties may become dominant, and multiphysics-scale problems arise. Design of bio-nanorobotic systems is a complex and need research on modeling and simulation of physical, chemical, mechanical and biological systems at the nanoscale. These include techniques that mainly cover the range of multidisciplines from atomistic mechanics to continuum mechanics: quantum mechanics, multi-particle simulation, molecular simulation, continuum-based models, stochastic methods, and nanomechanics. Approaches that make use of more than one such technique and focus on their integration play an important role in this effort.

Current proposed multiscale models for non-organic materials do not integrate the hierarchical design and focus mainly in the multiscale modeling of single components, such as carbon nanotubes (CNTs) [17] or electronic materials [18]. Some multiphysics models have been proposed but the proposed models does not take into consideration all physical interactions, such as quantum mechanics (QM), nanomechanics (NM) [19], or multiphysics parametrization in order to ensure connection

[1] http://www.ddl.unimi.it/vega/i

between length scales (from atomistic to macroscopic) [20]. Computational methods based on molecular dynamics (MD) are of great interest for investigations of structure, function, and dynamics of structural biology from the atomic level to the macromolecular level [21]. The fundamental issue is to integrate models addressing phenomena from the level of individual gene and cell features through tissue and organ models. Recent approaches are combining multiscale modeling of macromolecular conformational changes to concepts from rigidity and elastic network theory [22]. A good example is given in [23] where the structural dynamics of the lac repressor-DNA complex is described by all-atom molecular dynamics and continuous elastic ribbon. Still, due to their size, biomolecular systems characterized by computational methods like molecular dynamics, are often beyond the reach of real-time feedback for control.

The objective of our work is to demonstrate a framework that couple multiscale, multiphysics and virtual reality advanced techniques for the design and modeling of bio-nanorobotic systems. A three-level modeling paradigm was developed for this purpose. Quantum mechanics (QM) for designed components parametrization, molecular dynamics (MD) for mechanical properties characterization and continuum-model (CM) for system simulation.

Virtual Reality Interfaces

The application of advanced virtual reality (VR) techniques in bio-nanotechnology allowed to study, simulate, visualize and interact with molecular biology, genomics, proteomics, structural and computational biology. As far we know, only basic molecular elements have been studied until now. As a perspective of these developments, VR-based technology could be used for the biomolecular robot design (see Fig. 2.3). The multiscale prototyping simulator that is capable of supporting VR devices such as stereo glasses, 3D trackers, dataglove and force-feedback (haptic) devices is of great interest. This development effort will permit operators to build models, combine atomic and volumetric data, and perform manipulation, assembly, steering forces interactively within a single computational environment.

Haptic Interaction

In this section, we describe our design and implementation of a communication scheme connecting a force feedback device, visualization program, and simulation engine. The incorporation of virtual reality into molecular science has only just begun. Several types of software for displaying 3D models of bio-molecules have been developed. Most of these programs use Virtual Reality Modeling Language (VRML). Few attempts to use force feedback have been investigated. Force feedback technology for protein-ligand docking have been used, i.e., local-minima force field minimization using haptic interaction [24, 25], steered molecular dynamics for

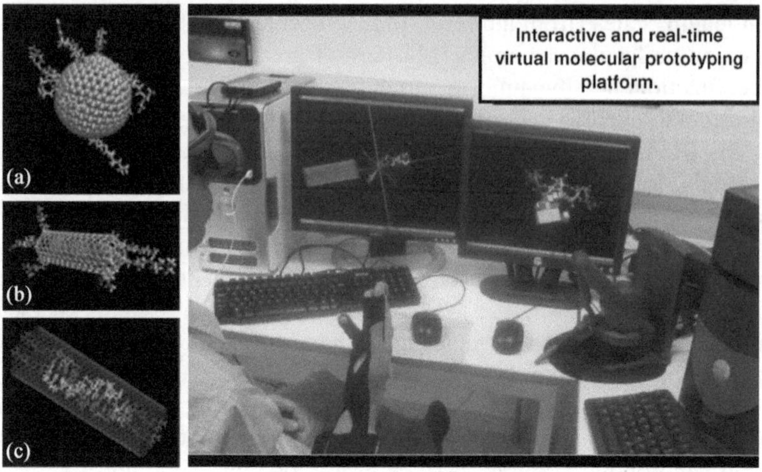

(d)

Fig. 2.3 *Left*: Pictures (**a**)–(**c**) show some bio-nanorobotic structure designs. *Right*: Picture (**d**) shows the experimental interactive simulation platform using virtual reality interfaces. In the virtual molecular dynamics (VMD) environment, the user applies forces to bio-nanorobotic structures in the simulation via a force-feedback haptic interface while manipulation is performed through a virtual hand. The headtracker is mounted on a pair of shutter glasses for operator immersion

protein stretching [26, 27] or feeling attractive and repulsive forces when screening molecules [28, 29]. The Interactive Molecular Dynamics (IMD) [14] permits manipulation of molecules in molecular dynamics simulations with real-time force feedback and graphical display. It provides interactive steering molecular dynamics calculations by adding user-specified external forces into the computation on the fly. Steering implies that the user is able to 'tug' an atom or a group of atoms towards a desired target position. Communication is achieved through an efficient socket connection between the visualization program (VMD) and a molecular dynamics program (NAMD) running on single or multiple machines.

A natural force feedback interface for molecular steering is provided by an haptic device (SensAble Technologies PHANToM Desktop interface). The physical forces generated by the user's hand are translated into forces on simulated atoms. Three scaling parameters define the relationship between the virtual haptic environment and the simulation environment.

- α, is the ratio of wall clock time required for a molecular dynamics timestep to the physical simulation time, and is presently at least 10^{12}.
- β, is the amount by which forces applied by the user are scaled before being applied to atoms in the simulation.
- γ, is the factor that scaled up atomic coordinates to be perceptible to humans.

These parameters have certain limitations. First, α and β can be set to any value the user desires; however, the value of β has important implications for the sensitivity and responsiveness of the IMD module. Second, β is limited by the workspace of the haptic device, although one could in practice focus on a small part of the simulation region in order to increase the spatial resolution of the haptic interface.

Let us to derive a model rendering the haptic force exerted on the user's hand. In the user-haptic coordinate system, let F_{haptic} be the force exerted by user's hand on the haptic device. The particle experiences a force β. F_{haptic}, taking into account the scaling of applied forces.

We can relate the dynamics of the particle as perceived in the haptic environment to the dynamics of the particle in the simulation. Let x and t be the space and time coordinate, respectively, of the particle in the haptic environment, and χ and τ be the corresponding simulation variables. The dynamics of an otherwise free particle in the simulation are given by

$$m\frac{d^2\chi}{d\tau^2} = \beta F_{haptic} = m\frac{\alpha^2}{\gamma}\frac{d^2x}{dt^2}. \tag{2.1}$$

The user interprets the motion of the particle in terms of F_{haptic} and an effective mass m^*, defined by

$$F_{haptic} = m^*\frac{d^2x}{dt^2}. \tag{2.2}$$

Combining (2.1) and (2.2), we arrive at an expression for the effective mass in terms of the actual mass of the particle and the scaling parameters:

$$m^* = m\frac{\alpha^2}{\gamma\beta}. \tag{2.3}$$

Now, consider the special case in which the applied force is proportional to the distance of the particle in the virtual environment from a restraint point that moves with the haptic controller. In this case $F_{haptic} = -kx$, where k is the spring constant defining the strength of the restraint. The atom experiences an external force $\beta kx = k\beta\gamma\chi$. If the particle is also subject to a potential U due to the presence of other atoms in the system, the dynamics of the particle will be given by

$$m\frac{d^2\chi}{d\tau^2} = -k\beta\chi\gamma - \frac{dU}{d\chi}. \tag{2.4}$$

Equations (2.3) and (2.4) together describe the interplay of the scaling parameters in determining the environment experienced by the user in IMD module. In order for the system to respond to applied forces, m^* must be reasonably small, i.e. a few kilograms, since the maximum force that can be exerted by modern haptic devices is ~1–10 N. As illustration, Fig. 2.4 demonstrates DNA molecule encapsulation into a carbon nanotube performed by the operator through force feedback interface. As the nanotube-DNA interaction experience strong attractive force at near distance (few nanometers), the operator feels unstable insertion forces and guides haptically the molecule along the nanotube.

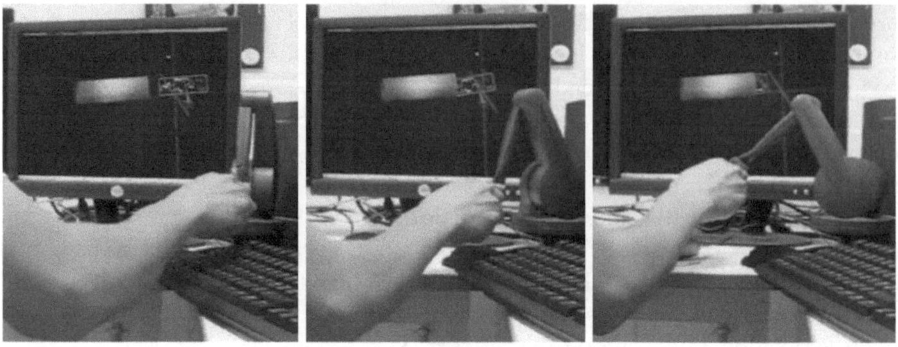

(a) (b) (c)

Fig. 2.4 DNA encapsulation in the carbon nanotube using operator force feedback

Tracking System and Input Devices

Virtual Environments (VE) offer a new potential by a more intensive experience of 3D spatial information and natural forms of interaction, to increase the understanding of dynamic spatial relationships. This intensive exposure of scientists to simulation models and data is achieved by representing these as 3D artificial worlds that are suitable for navigation, interaction, and quantification. Intuitive and natural interactions are key aspects of virtual nanoassembly applications.

Intuitive virtual hand interactions with parts, subassemblies and tools is provided by a CyberGlove dataglove (Ascension Technologies CyberGlove18 interface). The 18-sensor data glove features two bend sensors on each finger, four abduction sensors, plus sensors measuring thumb crossover, palm arch, wrist flexion, and wrist abduction. Connected to a 3D tracking device (tracker sensor mounted close to the wrist), it provide the CyberGlove motion capture data glove system with absolute x, y, z, yaw, pitch, roll position and orientation information. A complete tracking system includes one transmitter, one receiver, a system electronics unit with RS-232 serial port. The dataglove interface allows virtual hand representation for manual selection, manipulation and grasping schemes for virtual nanoassembly tasks. Proper registration and calibration of the tracking system is very important. Registration of the tracking system means that we define a transformation from the original tracker coordinate system to a known world coordinate system. Calibration of the tracking system should minimize the tracking errors. Research work is under investigation in order to define grasping patterns based on the abstraction of various mechanic shapes, kinematics model of virtual hand is built upon the anatomic structure of the human hand to guarantee motion realism, and real-time collision detection scheme is used to identify whether the current finger configuration matches a specific manipulation and grasping pattern. Navigation is provided to the user via uniform motion tracking of head interface with immersive displays (stereo glasses). The headtracker is mounted on a pair of LCD shutter glasses. The tracking system enables dynamic spatial measurements of the tracking sensors. The RWB is equipped with a Polhemus Fastrak system, consisting of a control unit, an electromagnetic field emitter and one or more tracking sensors. The control unit is connected via a serial port. The system is able to determine both spatial position (XYZ) and orientation (AER: azimuth, elevation, roll) of each of the sensors. This results in a total of six degrees of freedom (6 DOF) per sensor.

Interaction Virtual Environment

The user interface with its interaction methods plays a crucial role in nanorobotic design. A virtual environment provides the user interface for VR applications. By using spatial input devices, the user can interact with this computer-generated 3D Virtual Molecular Dynamics environment. In the design of the user interface of our system we concentrated on the user's tasks in scientific visualization and the specific characteristics of VR on the multiscale modeling tools. The designer can choose the desired organic and non-organic elements from an homemade "library bio-nano components". The library is composed of various organic (peptides, proteins, acids) and non-organic (carbon nanotubes, nanoparticles, polymers). The bio-nano components are accessible via a 3-D graphical user interface developed through the VMD environment (see Fig. 2.5). The attachment phase between the different links, structure or peptides with the CNTs can be initiated through 3D virtual manipulation and stereo visualization. The user performs manually the atomistic connections using the data-glove interface or the force feedback haptic interface.

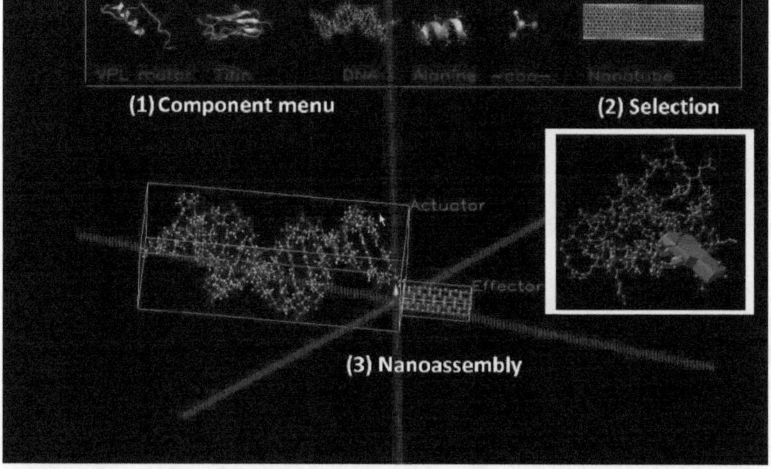

Fig. 2.5 Graphical user interface: bio-nanorobot components are accessible via a 3-D graphical user interface developed through the VMD environment. Illustration task: (*1*) Choice of the components in the library (proteins, biomotors, carbon nanotubes, etc.), (*2*) molecule handling and displacement to a desired location and (*3*) nanoassembly task between a molecule and carbon nanotube

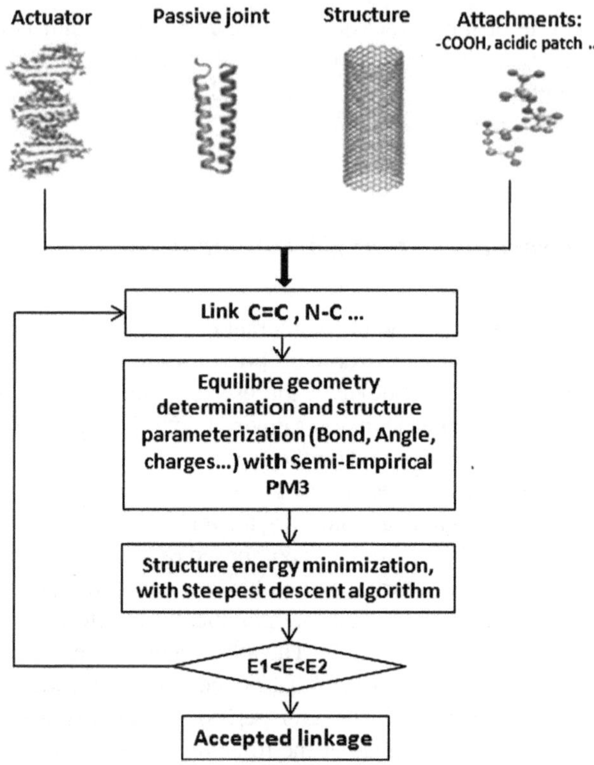

Fig. 2.6 Theoretical and computational nanoassembly methodology

Here we describe the proposed assembly methodology using the described CAD simulation platform. It involves the assembly of functionally stable bio-nano components into complex assemblies. The nanoassembly approach is based on hierarchical theoretical and computational assembly methods depicted on the flowchart shown in Fig. 2.6.

1. The designer can choose the desired organic and non-organic elements from an homemade component database (see Chap. 3). The database is composed of various nanorobotic components such as elastic proteins, carbon structures, protein-based actuators or molecular attachments. The different mechanical, dynamics and kinetics properties of the database structures have been previously characterized. The nanorobotic components are accessible via a 3-D graphical user interface (GUI) developed through the VMD environment.
2. The attachment phase between the different links, structure or proteins can then be initiated through 3D virtual manipulation and stereo visualization. The user performs atomistic connections, i.e. carbon-carbon (C–C), carbon-hydrogen (C–N) bonds in order to create new residues.
3. Such residues are then parameterized through QD simulation in terms of bond length, dihedral angles and Coulomb charges at around the equilibrium state. We use semi-empirical PM3 or Hartree-Fock molecular orbital models with SPARTAN and GAMESS tools.

4. The parameterized attachment is then defined as new residue in CHARMM27 force field. The assembled total nanorobot structure is energetically minimized through the steepest descendent algorithm (SDA) with the respect to two energetic barriers (El and E2) corresponding to the energies of acceptable robot geometry.
5. Thus, the program prevents the user from applying the tools in way that is not chemically reasonable.

Multiscale Modeling of Bio-Nano Robotic Structures

Understanding the relation between molecular behavior and system/device response, requires the development of modeling and simulation tools, which can analyze phenomena spanning the wide spectrum of length and time scales. This cannot be accomplished with one theory or simulation method. We have developed a design and modeling platform that dynamically couple all frameworks. We use several physics for particular length- and time-scales to characterize and treat aspects of nanostructure phenomena that operate only over those scales (Fig. 2.7). In the following we present the main characteristics of these methods and how they are combined within multiscale modeling strategies.

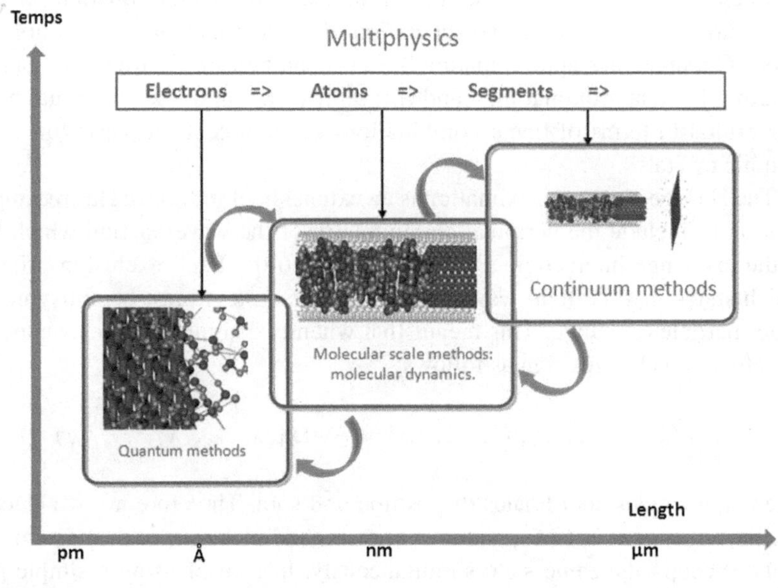

Fig. 2.7 Multiscale modeling of bio-nanorobotics structures

Quantum and Semi-empirical Methods

1. *Quantum Mechanics (QM)*: A wide variety of different procedures or "models" have been developed to calculate molecular structure and energetics among other properties. Theses have generally been broken down into two categories, quantum chemical models, which ultimately derive from the Schrödinger equation, and molecular mechanics models. In atomic units, the electronic Schrödinger equation is

$$\hat{H}\Psi(r;R) = \varepsilon\Psi(r;R) \tag{2.5}$$

where r and R denotes electronic and nuclear degrees of freedom and $\hat{H}$ denotes the Hamiltonian operator. The term $\hat{H}$ describes both the kinetic energies of the nuclei and electrons which makeup the molecule, as well as the electrostatic interactions felt between the nucleic and electrons. The Hamiltonian is given by

$$\hat{H} = -\frac{1}{2}\sum_i \nabla_i^2 - \sum_{r,i}\frac{Z_A}{r_{Ai}} - \sum_i \frac{Z_A Z_B}{R_{AB}} + \sum_{i>j}\frac{1}{r_{ij}} \tag{2.6}$$

where A and B design nuclei, and i and j electrons, Z are the atomic numbers, and atomic units are used in (2.5).

2. *Hartree-Fock Molecular Orbital Models*: Although the Schrödinger equation may easily be written down for many electron atoms as well as for molecules, it cannot be solved. Approximation must be made. So called Hartree-Fock molecular orbital models [30], start from the Schrodinger equation and then make three approximations: (i) separation of nuclear and electron motions (the "Born-Oppenheimer approximation"), (ii) separation of electrons motions (the "Hartree-Fock approximation") and (iii) representation of the individual molecular orbital in terms of linear combinations of atom-centered basis function or "atomic orbitals".

   The Hartree-Fock approximation is an extension of the above Hartree approximation to include the permutation symmetry of the wavefunction which leads to the exchange interaction. Exchange is due to the Pauli exclusion principle, which states that the total wavefunction for the system must be antisymmetric under particle exchange. This means that when two arguments are swapped the wavefunction changes sign as follows:

$$\Psi(x_1, x_2, \ldots, x_i, x_j, \ldots, x_N) = -\Psi(x_1, x_2, \ldots, x_j, \ldots, x_N)$$

   where x_i includes coordinates of position and spin. Therefore no two electrons can have the same set of quantum numbers, and electrons with the same spin cannot occupy the same state simultaneously. Instead of using a simple product form of the wavefunction, a Slater determinant wavefunction which satisfies antisymmetry is used. This can be represented in matrix notation as

$$D = \begin{vmatrix} \psi_1(1) & \psi_2(1) & \psi_3(1) & \cdots & \psi_n(1) \\ \psi_1(2) & \psi_1(2) & \psi_1(2) & \cdots & \psi_n(2) \\ \cdots & \cdots & \cdots & \cdots & \cdots \\ \psi_1(n) & \psi_2(n) & \psi_3(n) & \cdots & \psi_n(n) \end{vmatrix} \tag{2.7}$$

where $\psi_i(x_j)$ are the one-electron wavefunctions.

Following exactly the same method of minimizing the expectation value of the Hamiltonian $\hat{H}$ with respect to the one-electron wavefunctions as was used in the derivation of the Hartree equations, results in the following set of one-electron equations, the Hartree-Fock equations;

$$\epsilon_i \psi_i(r) = \left(-\frac{1}{2}\nabla^2 + U_{ion}(r)\right)\psi_i(r) + \sum_j \int dr' \frac{|\psi_j(r')|^2}{|r-r'|}\psi_j(r)$$
$$- \sum_j \delta_{s_i s_j} \int dr' \frac{\psi_j^*(r')\psi_i(r')}{|r-r'|}\psi_j(r)$$

where s_i labels the spin of particle i.

3. *Semi-Empirical Molecular Orbital Models*: The principal disadvantage of Hartree-Fock models is their cost [31]. It is possible to introduce further approximations in order to significantly reduce cost while still retaining the underlying quantum mechanical formalism. "Semi-empirical" molecular orbital models follow in a straightforward way from Hartree-Fock models: elimination of overlap between functions on different atoms, restriction to a "minimal valence basis set" of atomic functions and introduction of adjustable parameters to reproduce specific experimental data. The Stewart's Parametric Method Number 3 (PM3) semi-empirical model is particularly attractive for equilibrium and transition-state geometry determinations for large molecules where the costs of Hartree-Fock models may be prohibitive.
4. *Molecular Mechanics Models*: The alternative to quantum chemical models are molecular mechanics models [32]. These do not start from the Schrödinger equation, but rather from a simple but "chemically reasonable" picture of molecular structure, a so-called "force field". In this picture, just as with a Lewis structure, molecules are made up of atoms (as opposed to nuclei and electrons), some of which are connected ("Bonded"). Both crowding ("Van der Waals") and charges ("Coulombic") interactions between atoms are then considered, and atom position are adjusted to best match known structural data (bond lengths and angles). Molecular mechanics is much simpler than solving the Schrödinger equation.

   The PM3 semi-empirical model is suitable for identifying conformational minima, and for determining the geometries (bond lengths, angles and charges) of these minima.
5. *Semi-empirical Parameter Generation*: When attaching a residue termini to another one, such for example attaching leucine termini to graphite layer, a novel

(a) (b)

Fig. 2.8 (**a**) Leucine-ARM residues linkage and (**b**) linkage equilibrium geometry at ground state with Semi-Empirical PM3

nonstandard group become created (see Fig. 2.8(a)), the first task is to determine exactly which parameters are already known, and which will need to be developed from scratch. The unparameterized part of the new residue is the connection between the azote and the carbon. The connection between the azote of LEU and the carbonyl carbon of the ARM residue is called a thioester bond. Leucine-ARM linkage is defined as a new residue defined as a new residue in the CHARMM27 topology [33] file with the name LARM. In order to use this LARM residue in MD simulations, we need to make a topology file entry for it as well as develop a complete set of parameters describing it. The unparameterized part of the LARM residue is the connection between the leucine and the carbon.

To actually determine the optimum parameter values for equilibrium bond lengths, angles, and to determine the energetic barriers associated with them, we need to use quantum chemistry. For our study, we use SPARTAN to generate these values. We calculate the values semi-empirically. With semi-empirical quantum mechanics calculation, we optimize the geometry of the structure as shown in Fig. 3.16(b), calculate electrostatic potential (ESP) charges and vibrational frequencies. The LARM linkage, which correspond to C–N bond, calculated by SPARTAN semi-empirical PM3 methods is 1.449 Å and C–N–C angle is 123.25°. The LARM linkage parameters (angles, bonds lengths, charges) will be add to CHARMM27 force field to be used in molecular dynamics level.

Molecular Dynamics

We carry out computer simulations in the hope of understanding the properties of assemblies of molecules in terms of their structure and the microscopic interactions between them. This serves as a complement to conventional experiments, enabling us to learn something new, something that cannot be found out in other ways. The two main families of simulation technique are molecular dynamics (MD) and Monte Carlo (MC) [33]; additionally, there is a whole range of hybrid techniques which combine features from both. In this lecture we shall concentrate on MD. The obvious advantage of MD over MC is that it gives a route to dynamical properties of the system: transport coefficients, time-dependent responses to perturbations, rheological properties and spectra.

To predict the dynamic performance of a designed nanodevice (i.e., energy and force calculation), we are performing Molecular Dynamics (MD) Simulations based on the calculation of the free energy that is released during the transition from native to fusogenic state. We used MD software called NAMD (Chemistry at Harvard Molecular Mechanics) [15]. In MD, the feasibility of a particular conformation of the biomolecule in question is dictated by the energy constraints. Hence, a transition from one given state to another must be energetically favorable, unless there is an external impetus that helps the molecule overcome the energy barrier. When a macromolecule changes conformation, the interactions of its individual atoms with each other-as well as with the solvent compose a very complex force system. The NAMD energy function is given:

$$U(\overrightarrow{r1}, \overrightarrow{r2}, \ldots, \overrightarrow{r_N}) = U(\overrightarrow{R}). \tag{2.8}$$

The equation used to determine the force on each atom as:

$$m_i \frac{d^2 \overrightarrow{ri}}{dt^2} = \overrightarrow{F_i} = -\overrightarrow{\nabla} U(\overrightarrow{R}). \tag{2.9}$$

Newton's equation represents a set of N second order differential equations which are solved numerically at discrete time steps to determine the trajectory of each atom.

$$\overrightarrow{r_i}(t+\Delta t) = 2\overrightarrow{r_i}(t) - \overrightarrow{r_i}(t-\Delta t) + \frac{\Delta t^2}{m_i}\overrightarrow{F_i}(t). \tag{2.10}$$

The total energy is given by:

$$U(\overrightarrow{R}) = U_b + U_\theta + U_\omega + E_{vdw} + E_{elec}. \tag{2.11}$$

The individual components of the total energy U are the bond potential U_b, the bond angle potential U_θ, the dihedral angle potential U_ω, the Van der Waals Potential U_{vdw}, the electrostatic potential function U_{elec}. The functional forms of these potentials are as follows: Bond potential is given by

$$U_b = \sum_{bonds} k_i^{bond}(r_i - r_0)^2.$$

The variable r is an instantaneous bond length due to an extension or contraction of a bond and r_0 is its original length. Similarly, the bond angle potential is given by

$$U_\theta = \sum_{angles} k_i^{angle}(\theta_i - \theta_0)^2. \tag{2.12}$$

Angle θ is the deviation from the equilibrium value θ of the bond angle for a given set of atoms. Under normal temperature conditions, these two potential values are

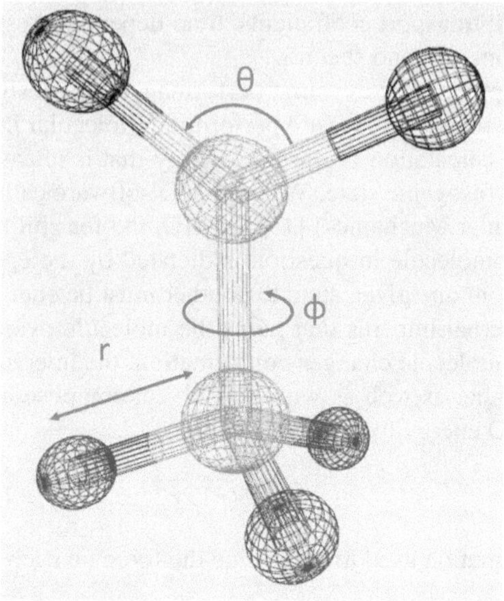

Fig. 2.9 Internal coordinates for bonded interactions: r governs bond stretching; θ represents the bond angle term; ϕ gives the dihedral angle

very small, as the bond lengths and angles do not change appreciably. However, there is rotation around certain bonds (torsion or bond twist). The dihedral angle or torsion potential is thus formulated, which is given by

$$U_{\omega} = \sum_{dihedrals} k_i^{dihe}[1 + \cos(n_i \phi_i + \delta_i)]. \tag{2.13}$$

This term is a four-atom term based on the dihedral angle ϕ about an axis defined by the middle pair of atoms. The electrodynamic effect of Van der Waals forces is included by using the Lennard-Jones potential as:

$$U_{vdw} = \sum_i \sum_{j \neq i} 4\epsilon_{ij} \left[\left(\frac{\sigma_{ij}}{r_{ij}} \right)^{12} - \left(\frac{\sigma_{ij}}{r_{ij}} \right)^{6} \right]. \tag{2.14}$$

In the above equation, ϵ_{ij} represents the energy of the minimum (deepest) point on Van der Waals curve for the atom pair $i - j$, σ_{ij} is the separation distance between the atom pair $i - j$ at the energy minima, and r_{ij} is the current actual distance between atom i and atom j. The attractive forces between two proximal atoms are proportional to r_{ij}^{-6} and the repulsive forces due to their nuclear repulsions are proportional to r_{ij}^{-12}. The other important factor in nonbonded interaction is the electrostatic interaction, which for atoms i and j, is given by:

$$U_{elec} = \sum_i \sum_{j \neq i} \frac{q_i q_j}{\epsilon r_{ij}}. \tag{2.15}$$

Combined Molecular Dynamics and Continuum Approach

The main goal of molecular dynamics studies is to understand the relation between molecular behavior and nanorobot response. Molecular dynamics computing time may take a very long time, thus evaluation of the controller design can not be achieved through realtime simulation. Some nanorobot elements, such as Immunoglobulin, DNA and Fibronectin has been described as a molecule that acts like a molecular spring and different corresponding models can be adopted. To overcome the long time computing problem, the proteins spring was modeled by the Worm Like Chain (WLC) or Hookian model in molecular dynamics. The Hookian and WLC model was implemented as TCL script, and communicate in realtime with NAMD program during simulations. In order to compare the performances of quantum mechanics-molecular dynamics (QM-MD) and quantum mechanics-molecular dynamics-continuum mechanics (QM-MD-CM), we performed several calculations. It can be noticed that for acceptable results of simulation, the QM-MD-CM reduces the complexity of nanorobotic structures (decrease of number of atoms) to be simulated for smaller computational time (around few milliseconds) and smaller simulation times (nanosecond order), those results will be detailed in the next chapter. We adopt WLC model for passive joints-based proteins Fibronectin Fn-III_{10}, Immunoglobulin I27, Tropocollagen, silk, DNA and Tropomyosin, that describes a molecular chain as deformable continuum or rod of given persistance length, A, which is a measure of the molecule's stiffness. The relationship between the end-to-end length (z) and the external force (f) are given by:

$$f = \frac{KT}{A}\left[\frac{1}{4(1-\frac{Z}{L})^2} - \frac{1}{4} + \frac{z}{L}\right] \tag{2.16}$$

where k is the Boltzman constant, T is the temperature, and L is contour length (the length of fully extended WLC chain). For our simulations the persistence length is settled to $A = 0.4$ nm (see [35]). In order to determine the value of the protein length L, we summed the contour lengths of all the protein domains when considered in its fully extended state. It results that the contour length is proportional to the number of amino acids [36]. The corresponding external force increases linearly as a Hookes for α-helix deca-alanine, repressor of primer protein and fibrillin protein. The extension of the Hookian spring is proportional to the force applied at its end such as:

$$f = \left(\frac{k_B T}{A}\right)\left(\frac{x}{L}\right). \tag{2.17}$$

Mechanical Characterization by Steered Molecular Dynamics

Biological springs can be classified as active mechano-chemical devices that store the energy of conformation from various bonded and non-bonded interactions as potential energy. To characterize the mechanochemical energy conversion and to study

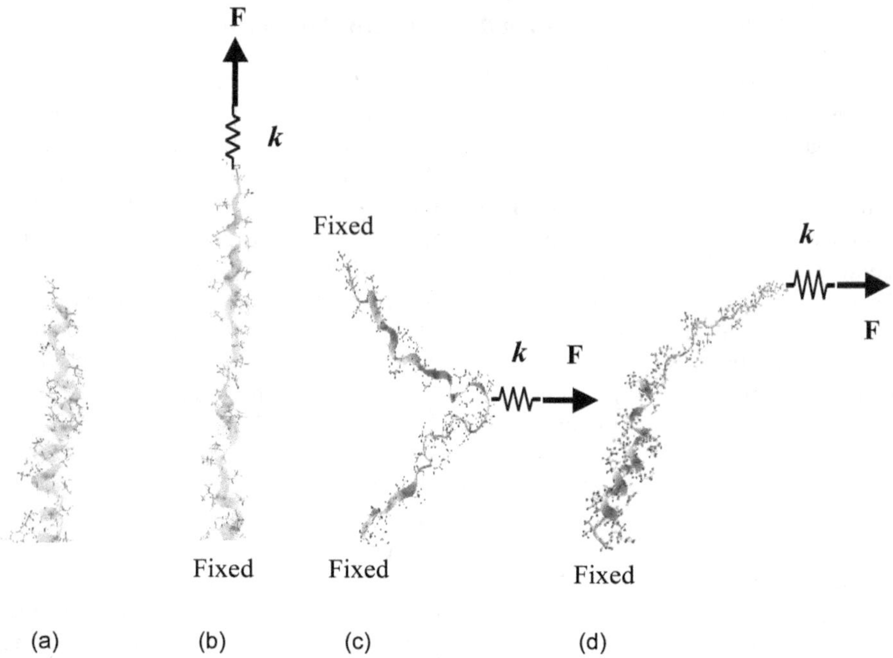

Fig. 2.10 Schematic diagrams of the various deformations studied with the attachment points used for mechanical unfolding experiments. *The arrows* indicate the positions where the stretching constraints were applied and the direction of stretching. (**a**) Folded configuration of helical protein, (**b**) longitudinal stretching, (**c**) lateral bending and (**d**) lateral shearing. The moving guiding potential used in the stretching simulations is represented by a spring k which is pulled with a constant velocity v

the elasto-mechanical properties of these biological springs external are applied using steered molecular dynamics. To study some mechanical proprieties various deformations properties of mechanical proteins can be simulated when subjected to external stretching. First, (i) in longitudinal stretching (Fig. 2.10(b)), we fix one end of the molecule (the C-atom of the N-terminus) at the origin and constrain the other end (the capping nitrogen atom of the C-terminus) to move only along the longitudinal axis, thereby removing the irrelevant degrees of freedom, i.e., removing the overall translation and rotation. Then, (ii) in lateral bending (Fig. 2.10(c)), the C- and N-terminals are fixed and the α carbon at the center is harmonically pulled in lateral direction. Finally, (iii) in lateral shearing (Fig. 2.10(d)), the C-terminal is held fixed and the N-terminal is harmonically pulled in lateral direction. The moving guiding potential:

$$u(r,\lambda) = \frac{k}{2}[\xi(r) - \lambda]^2 \tag{2.18}$$

is added to control the end-to-end distance x which is a function of the 3N-dimensional position r of the system. The moving guiding potential used in the pulling simulations is represented by a spring which is connected to the C-terminus and pulled with a constant velocity v. For each protein, the values of the spring constant k and stretching velocity v were chosen to correspond pulling with a stiff spring in the drift regime. the external force exerted on the system can be expressed as

$$F = K(x_0 + vt - x). \tag{2.19}$$

This force corresponds to the ligand being pulled by a harmonic spring of stiffness K with its end moving with velocity v. Alternatively, a fixed restraint point at a distance much larger than the length of the unbinding pathway may be chosen. In this case, the end of the spring does not move and its stiffness is linearly increased with time [37], i.e., $K = \alpha t$, and the force is

$$F = \alpha t(x_0 - x). \tag{2.20}$$

Control Parameters Coupled to Molecular Dynamics

pH Control

Numerous protein-based nanodevices show a spontaneous "opening" and "closing" mechanism upon temperature, pH, light, ionic strength and electron density modulation and can be considered as a basic molecular motor, nanogripper or nanosensor capable of generating mechanical force during its 3D conformational transition. These control parameters have been included in the molecular dynamics algorithms in order to prototype controlled mechanical motion. As illustration, Fig. 2.11 shows the conformation control of a viral protein motor (VPL) when considering the pH variation. It changes the energetics of the outer (envelope glycoprotein) protein in such a way that there is a distinct conformational change in a part of it. Proteins

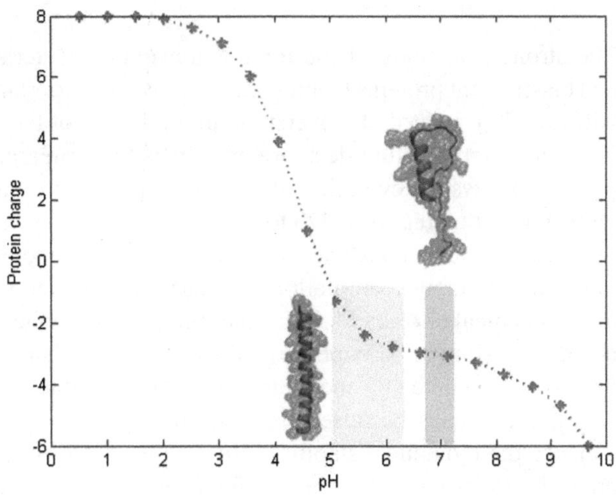

Fig. 2.11 Charge on amino acids side chains in viral protein with respect to the level of pH

contain many ionizable groups on the side chain of their amino acids. pH is the negative log of the hydrogen ion concentration in a solution such as $pH = -\log[H^+]$. For a neutral acidic pH value ($pH = 7$), the VPL protein is in a partially α-helical stranded coiled coil. The random coil regions are converted into well defined helices and an extension occurs at lower pH between 5 and 6.5. For this configuration, it is necessary to protonate the amino acid side chain of the protein. Because Molecular dynamics simulations are used to simulate the motion of molecules in order to gain a deeper understanding of the chemical reactions, fluid flow, phase transitions, and other physical phenomena due to molecular interactions. Rapidly increasing computational power has made MD simulation a powerful tool for studying the structure and dynamics of biologically important molecules. Taking into account all electrostatic interactions by using the Particle Mesh Ewald (PME) method, relatively long (2–3 ns) simulations with the explicit solvent water box can be carried out. We perform constant-pH molecular dynamics (MD) simulations with an implicit solvent and explicit treatment of protonation/deprotonation phenomena. The protonation state of each ionizable group in the molecule is allowed to change during the MD trajectory at predefined time intervals, dividing the whole simulation into a set of sub-trajectories. Each change in the protonation state of the molecule at the beginning of a new sub-trajectory is made with a probability determined by the electrostatic free energies of the available states, based on the last structure of the preceding sub-trajectory.

In Chap. 3, the influence of these control parameters will be discussed in detail through the MD study of different bio-nano components. As experimental evaluation for such optimum conditions is very time consuming and costly, computer-aided modeling and simulation studies are helpful to identify the optimum control parameters in an effective way.

Thermal Control of Nanodevices

The strong sensitivity of the force-extension characteristics of the DNA molecule and elastic joint proteins to temperature plays an important role in the control of their stiffness [21]. Indeed, the overstretching portions of the force-extension curves as a function of temperature decrease greatly as the temperature is increased. As example, [56] shows the evolution of the DNA spring constant K_{DNA} with respect to the temperature ranging from 273 to 450 K. In theory, the results show clearly the linear decrease of spring constant with respect to the temperature. Constant temperature molecular dynamics simulations are used to simulate the effect of temperature in the mechanical characteristics of biological elastic joint. Constant temperature control during simulation is performed by the use of canonical ensemble *NVT* where, moles (N), volume (V) and temperature (T) are conserved. NAMD use a stochastic coupling approach because it is easier to implement and the friction terms tend to enhance the dynamical stability. The Langevin equation is used in NAMD to generate the Boltzmann distribution for canonical (*NVT*) ensemble simulations. The generic Langevin equation is

$$M\dot{v} = F(r) - \gamma v + \sqrt{\frac{2\gamma k_B T}{M}} R(t) \tag{2.21}$$

where M is the mass, $v = \dot{r}$ is the velocity, F is the force, r is the position, γ is the friction coefficient, k_B is the Boltzmann constant, T is the temperature, and $R(t)$ is a univariate Gaussian random process. Coupling to the reservoir is modeled by adding the fluctuating (the last term) and dissipitive ($-\gamma v$ term) forces to the Newtonian equations of the motion. To integrate the Langevin equation, NAMD uses the Brüger-Brooks-Karplus (BBK) method, a naturel extension of the Verlet method for the Langevin equation. The position recurrence relation of the BBK method is

$$\begin{aligned} r_{n+1} = r_n &+ \frac{1-\gamma\Delta t/2}{1+\gamma\Delta t/2}(r_n - rn + 1) \\ &+ \frac{1}{1+\gamma\Delta t/2}\Delta t^2\left[M^{-1}F(r_n) + \sqrt{\frac{2\gamma k_B T}{\Delta M}} Z_n\right] \end{aligned} \tag{2.22}$$

where Z_n is a set of Gaussian random variables of zero mean and variance 1. The BBK integrator requires only one random number for each degree of freedom. The steady-state distribution generated by the BBK method has an error proportional to Δt^2, although the error in the time correlation function can have an error proportional to Δt.

3 Co-prototyping of Nanorobotic Structures

Carbon Nanotubes Based Nanodevice Co-prototyping

Generally, prototyping of nanorobotic devices and structures may be impractical due to very high cost or low performance requirements. Much of the costs of today's processes are the result of meticulous design prior to fabrication and extensive testing (e.g. defect detection) after fabrication. A solution would be to use hardware and software co-design methodologies focusing on the prediction of system performance or co-verification of system functionality Fig. 2.12, the design stage is subdivided into two main parallel steps (namely, computational simulation and experimentation) leading to the nanodevice prototyping stage. As molecular simulations allow running several nanodevice tests within a theoretical model, we can update this computational model from real nanodevice experimentation leading to realistic and low-cost testing and prototyping.

As illustration of this concept, we developed a co-prototyping methodology applicable to various multiwalled carbon nanostructures. Figure 2.13 outlines the main steps ranging from basic design specifications to optimized prototyping of the

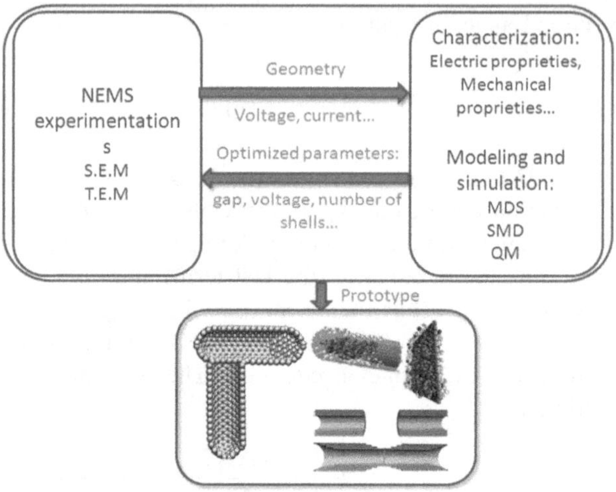

Fig. 2.12 Co-design and co-prototyping of based carbon nanotubes nanodevices

nanosystem. The first step in is the design of the nanosystem model by coupling theoretical data to carbon nanotube experimental measures.

The second phase, presents the nanodevice co-prototyping. The study and the characterization of the nanodevice is performed mainly by simulations based on realistic models. The model of the structure is updated interactively from the experimental. Simulations allow mainly three functions: characterize and explain the nanodevice behavior, to optimize the nanodevice (such as the determination of the configuration which correspond to a minimum of consumed electric potential) and finally to observe specific phenomena, which are not observable by experimentations (such as the rotation of the inner carbon nanotubes), this can be helpful for the design of the new nanodevices.

After the prototype has been validated, it is all prepared for the serial fabrication of the prototype.

Figure 2.1 illustrates the outline of co-prototyping method. In the proposed co-prototyping methodology, the prototype was intuitively designed in the above step as described by Fig. 2.13. Prototyping of the proposed nanodevice require several repetitive experimental tasks, because the hight cost of these kind of manipulation we use theoretical approaches to performs specific prototyping tasks. Prototyping flow begins with specification of designed system, by gathering the proprieties of the system from experimental platform, mainly from experimentation we consider the device dimension and the control parameters such as the actuation voltage. Prototyping the nanodevice by using simulations require system characterization and modeling. The designed system is characterized, electric proprieties are carried out, by using Quantum methods and atomistic moment method we calculate the charge redistribution along the nanodevice. We perform molecular dynamics simulations (MDS) to study the system characterization. As initial model used in MDS we generate the 3D system coordinate (xyz) by using dimensions gathered from experimentation. The we perform molecular dynamics simulation to characterize specific proprieties of the proposed system, such as the optimal actuation voltage consid-

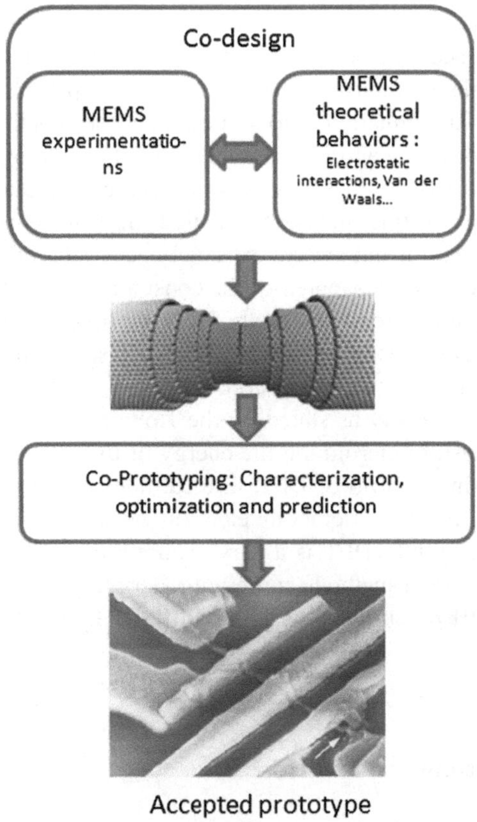

Fig. 2.13 Co-design of carbon nanotube based nanodevice

ering the number of shells in a based multiwalled carbon nanotube devices. The optimal actuation voltage for different configuration.

Simulation Methods

The methods of modeling the atomic interactions can roughly be divided into three categories: The first principles, or *ab-initio*, methods are the most rigorous ones. Only a few well controlled approximations to exact quantum mechanics are made. Semi-empirical methods contain more drastic approximations and may also contain empirically defined parameters. Empirical methods are a group of methods that are tuned to reproduce an empirically defined fitting set. The three categories, their drawbacks and benefits, and the most popular methods of each are briefly described in this section. The most rigorous approach to problems of condensed matter physics are the first principles methods. They do not employ any empirical parameters in the calculation. The results are based on quantum mechanics and well controlled approximations. The two most used first principles methods are the Hartree-Fock method and the density functional theory [38, 39]. The first principles methods are highly accurate but the system size that can be handled is restricted to tens, at most a couple of hundred atoms. The Hartree approach to the quantum mechanical many body wave function is to approximate the function by a product of single particle functions. The approximation is refined by adding Fermi-Dirac statistics by replacing the product of the wave functions by the Slater determinant. The method obtained with this inclusion is called the Hartree-Fock approach [30]. Even though the method is not always extremely accurate, it is much used. The accuracy can be enhanced but this increases the consumption of CPU-time [40]. In its most rigorous form,the Hartree-Fock method can be used on systems of tens of atoms. Density Functional Theory (DFT) [38, 39] is based on using the electron density $n(\vec{r})$ of the system as the basic variable. The ground state is completely described by the electron density as stated by the Kohn-Sham theorem [38, 39]. There are various schemes of determining the energy of the system from the electron density. In the most simple form of DFT, in local density approximation (LDA), the expressions based on a non-interacting electron gas at the local electron density of the real system. Currently DFT is a very accurate method and its accuracy can be enhanced by the use of methods combining Hartree-Fock and DFT description, for example, B3LYP1 functional description [40]. DFT can be used on systems of a few hundred atoms.

Semi-empirical Methods

Semi-empirical methods employ various approximations and may include experimentally fitted parameters. Their power with respect to the first principles methods lies in a smaller consumption of CPU-time but they must be employed more carefully to obtain reliable results. Semi-empirical methods include interaction schemes like the tight binding method (TB) [41] and various molecular orbital methods of computational chemistry [40]. The tight binding method avoids most of the heavy calculation of the *ab-initio* methods. There are several methods in this family. The TB-methods have in common a number of simplifications such as minimal basis sets compared to the first-principles methods and elimination of difficult integrals, that are either made small by mathematical transformations or used as parameters to be fit to experimental data. In principle, everything that can be calculated by the *ab-initio* methods can be calculated by an appropriate tight binding method and with less CPU-time. The keyword above is the 'appropriate': the generality and the transferability of a tight binding method depend on the approximations. Although valid for thousands of particles, the method used must be assessed case-specifically. In computational chemistry the methods similar to tight binding are called the neglect of diatomic differential overlap methods [40]. Also these methods simplify the involved integrals but retain the molecular orbital description. The group involves various methods such as AM12 and PM33 [40].

Empirical Methods

The category of empirical methods contains a wide set of parameterized classical force fields that reproduce more or less accurately the fitting set that is determined either from experimental data or from *ab-initio* simulations. Force-field methods have the benefit of being computationally simple and thus fast. They allow the simulation of large systems (up to hundreds of millions of atoms) over a greater time-span (up to microseconds) than the methods presented above. The simplest approximation for the interaction potential takes into account only two-particle interactions. Potentials like this are called pair potentials. The Lennard-Jones potential [42] and the Morse potential [43] are two very well known examples of pair potentials. Although pair potentials are not ideal for the study of the mechanical properties of materials, they are very much used because they are simple to implement and algorithms based on pair potentials are kind on computational resources. There are, however, some severe shortcomings which should be taken into account whenever pair potentials are used. For example, an interaction described by a pair potential model can depend only on the distance between the two particles. Thus the potential can not model directional bonding. Pair potentials often predict wrong vacancy formation energies and melting temperatures [42, 43]. The construction of an accurate force-field encounters often system and material based difficulties, and often even fundamental limitations. Difficult media are metals, alloys, semiconductors, and oxide-based insulators. In this thesis the emphasis is, of course, on carbon. Describing carbon accurately requires a lot of caution and is difficult because the 2s electrons participate in molecular bonding and together with the 2p electron orbitals hybridize to form a wide variety of potential bonding configurations that depend on the environment and on the ambient conditions. Even by itself, carbon can form structures as diverse as the isotropic diamond crystal or the anisotropic planar graphite and nanostructures. If the consideration is extended to organic molecules, even only hydrocarbons, the bonding variety, and at the same time the complexity of describing it accurately, becomes vast. If covalently bonding materials, like sp2-carbon, are simulated, the modeling of directional bonding is essential in order not to obtain outright wrong results. In general this means that the interaction potential used must include at least a three-body interaction term. Because carbon is the basis of all organic materials and because diamond and graphite have various applications in materials science, numerous attempts to create accurate classical interaction potentials have been made but only a few of the proposed schemes are widespread. For example, carbon interaction potentials developed by Tersoff in [44–46] and Brenner in [47] are widely used in the modeling of carbon. The Brenner model is considered more versatile because of additional conjugated bond description while the structures modeled by the Tersoff model suffer from an overbinding of radicals and the description of systems involving bonds that exhibit a mixture of sp3 and sp2 hybridization is inaccurate. Both models are short ranged and cover only the nearest neighbors to speed up the calculations. For description of graphite or multi-walled nanotubes, much longer range interactions are required. Various long-range extensions that are most often based on Lennard-Jones type additional term to describe the ʚ-bonding have been proposed. Stuart et al. [48] have aimed at maintaining the reactivity description of the Brenner [47] model while introducing the long-range interactions. Therefore, this model was chosen where long-range interactions are required. In general the force-field methods can give information on the structure and dynamics of a system, the total energies, entropies, free energies, and diffusive processes in the system. By their construction the methods are incapable of predicting any properties related to the electronic structure such as electrical conductivity, optical, or magnetic properties. In short, the classical force field methods are structural tools. The general accuracy of results obtained by classical force field methods is far from the level of accuracy obtained by *ab-initio* simulations but so are the feasible system size and attainable time scale as well.

In this work the emphasis is on studying carbon nanotube properties through classical means. The choice allows studying structures that have dimensions comparable to the experimentally observed ones. Classical molecular dynamics with AIREBO potential [48] are employed in the simulation. The potential used for this work is based on the reactive empirical bond-order potential of Brenner.

The reactive empirical bond-order (REBO) model is a widely-used theory to calculate the potential energy of covalent bonds. In 1988, based on works of Abell [49], Tersoff proposed a many-body energy function [44] parameterized for carbon [45] and multicomponent systems [46]. The expression of the total potential energy of system is a sum of nearest-neighbour pair interactions which depend not only on the coordination of atoms but also on their local atomic environment. A simple parameterized bond order function was used to describe chemical pair bonded interactions. This theory could realistically describe single, double and triple bond energies of carbon structures such as some hydrocarbons and diamond. A significant step was taken by Brenner for conjugated hydrocarbons and carbon system in 1990. He extended Tersoff's potential function by introducing two additional terms into the bond order to describe radical and conjugated bonds.

Because only nearest-neighbour interactions have been considered, the REBO function could be quickly evaluated. This computational efficiency could therefore be well adapted in numeric simulations [50]. In the research works concerning the mechanical properties of carbon nanotubes in past 20 years, REBO functions have been used to study the elastic properties [51] and the structural stabilities (defect, buckling, ultimate strength and bond breaking, ...) [52] of CNTs. An acceptable accuracy has been reported in some of these works, by comparing with the experimental, ab initio or tightbinding results. Despite numerous successful applications of the first-generation REBO potential functions, they suffer from several drawbacks. First, its form is too restrictive to simultaneously fit equilibrium distances, energies, and force constants for all types of carbon-carbon bonds. Second, the possibility of modeling processes involving energetic atomic collisions is limited because both Morse-type terms go to finite values as the distance between atoms decreases. Third, the absence of dispersion and the non-bonded repulsion term make the potential poorly suited for any system with significant intermolecular interaction. In the simulation of carbon nanotubes under bending strain, the REBO potential has been reported to lead direct inaccuracy at high level of deformation [53]. Finally, it has been reported by Pettifor and co-workers [54] that the neglect of a separate ð bond contribution leads to problems with the overbinding of radicals and a poor treatment of conjugacy. So as to resolve these problems, an adaptive intermolecular reactive bond order (AIREBO) function has been proposed by Stuart and co-workers [48] in 2000. In this theory, the entire system energy was given by the expression

$$U^p = U^{REBO} + U^{non\text{-}bonded} + U^{single\text{-}bond\text{-}torsion} \tag{2.23}$$

$U^{non\text{-}bonded}$ is the non-bonded interactions term (Lennard-Jones 12-6 potential function), and $E^{single\text{-}bond\text{-}torsion}$ represents the torsional interactions about single bonds. Both the repulsive and the attractive pair interaction functions have been modified to simultaneously fit bond properties that could not be fitted by the first-generation Brenner's function. This potential function can be well adapted to reactive systems.

In the REBO potential, the chemical binding energy U_b is a collection of that of individual atoms, in which only the nearest-neighbour interactions are considered.

$$U_b = \sum_{i=1}^{N} \sum_{\substack{j=1 \\ i=1}}^{N} \left[V^R(r_{i,j}) - b_{i,j} V^A(r_{i,j}) \right] \tag{2.24}$$

where $\varphi^R(r_{i,j})$ and $\varphi^A(r_{i,j})$ are the interatomic repulsion and attraction terms, r_{ij} is the distance between atoms i and j and b_{ij} is the bond order function between atoms i and j that is derived from Huckel electronic structure theory. All many-body effects are included in this function.

In the AIREBO potential, the dispersion and intermolecular repulsion interactions are modeled with a Lennard-Jones 12-6 potential function

$$U_{LJ} = 4\epsilon \left[\left(\frac{\sigma_{ij}}{r_{ji}} \right)^{12} - \left(\frac{\sigma_{ij}}{r_{ji}} \right)^{6} \right]. \tag{2.25}$$

The other new component of the AIREBO potential is a term dependent on dihedral angles. The original REBO potential lacked any torsional interactions about single bonds, reflecting its original focus on network solids such as diamond and small molecular fragments relevant to the chemical vapor deposition of diamond. With no barrier to rotation about single bonds, however, the original REBO potential is poorly suited for the simulation of saturated hydrocarbons larger than methane. As with the nonbonded interactions, the reactive nature of the potential requires some modifications to the conventional method of introducing torsional interactions. The usual form for the torsional potential is a cosine power series in the dihedral angle ω.

The total potential energy of system is

$$U^{tot} = U^{elec} + U^P \tag{2.26}$$

where U^{elec} is the electrostatic energy due to the extra charges on two carbon atoms and U^p is the AIREBO interatomic potential. The electrostatic energy can be calculated by the coulomb law: $E_{ij}^q = \frac{q_i q_j}{4\pi\varepsilon R_{ij}}$, where q_i and q_j are extra charges on carbon atoms, R_{ij} is the interatomic distance, and ε is the permittivity.

The charge distribution on the carbon nanotube is obtained by an atomistic moment method based on classical electrostatics theory [55]. In our simulations, we consider a multi walled carbon nanotube (MWCNT) with finite length. For simplicity, the nanotube is assumed to be located in an idealized electric field, i.e., the voltage on the on the nanotube surface is V_0. If we consider a carbon nanotube with n atoms, the potential at an arbitrary atomic position is given by [55]:

$$V(r_i) = \sum_{j=i}^{n} \frac{q_j}{4\pi\varepsilon_0 |r_i - r_j|} \tag{2.27}$$

where V is the electric potential, q_j the point charge on the atoms, r_i represents the location of the charged atom, and ε_0 is the permittivity of vacuum. Because the equipotential status of the nanotube surface, n equations can be written in a matrix form, i.e.,

$$[A]\{q\} = V \tag{2.28}$$

where $\{q\}$, V are the charge vector and the potential vector, respectively, and $[A]$ is the $n * n$ order matrix. Equation (2.28) is no longer applicable. For calculating the diagonal terms of the coefficient matrix, we need to recall the original assumption that the charge is uniformly distributed on a triangular surface area surrounding the carbon atom, as shown in Fig. 2.14. For simplicity, we treat the actual three-dimensional curve surface approximately as a flat triangle. The area of the triangle is $s = \frac{3\sqrt{3}b^2}{4}$, where $b = 0.142$ nm is the carbon bond-length in a carbon nanotube. The charge density on this triangular area can be written as $d = \frac{q_i}{s}$. The electric potential at the center of the triangle, i.e. the atom i, is resulted from the uniformly distributed charge and can be expressed as [55]

$$V(r_i)_i = \iint_s \frac{q_i/s}{4\pi\epsilon_0\sqrt{\xi + \eta}} d\xi\, d\eta. \tag{2.29}$$

The diagonal term extracted from (2.29) is

$$a_{ij} = \frac{1}{4\pi\epsilon_0 s} \iint_s \frac{1}{\sqrt{\xi + \eta}} d\xi\, d\eta \tag{2.30}$$

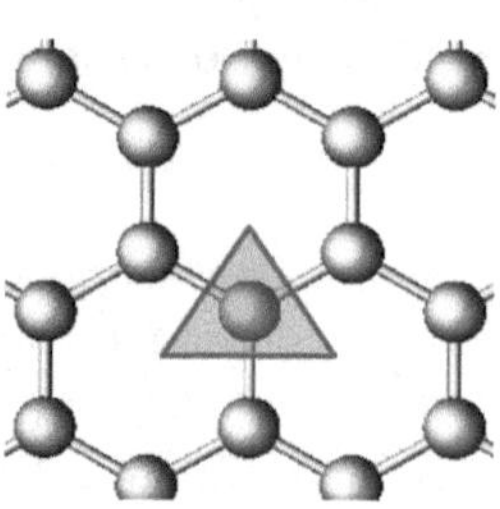

Fig. 2.14 Electric charge distributed on a triangular area surrounding a carbon atom

which has a closed form expression as

$$a_{ij} = \frac{b\sqrt{3}\ln(2+\sqrt{3})}{4\pi\epsilon_0 s}. \tag{2.31}$$

4 Conclusion

The objective of this chapter was to interface multiscale simulation tools with real-time truly virtual reality techniques for bio-nano robotic prototyping at the design stage. Molecular modeling is advocated here as a key methodology for research and development since it predicts the nanoscale interaction of unfamiliar combinations of biological and inorganic materials, and evaluates strategies for designing nanostructures for nanotechnological uses. The ability to interact with a computer-generated object in the same manner that a person would interact with a physical object—to investigate its structure by simply moving around it, to change its position by grabbing the object and moving the hand in space, without such artificial devices as computer mice. The design and prototyping methodology is illustrated in this chapter through the proposition of two case studies.

The first one involves multiscale modeling tools (quantum mechanics, molecular dynamics, continuum mechanics) coupled to virtual reality advanced techniques. In order to design and evaluate the characteristics of molecular robots, we proposed nanophysics-based simulation coupled to multimodality perception and real time computation. We have introduced a collection of utilities and programming libraries aimed at enhanced sampling and multiscale modeling applications in structural biology and nanotechnology. The tool set interfaces with the standard quantum and semi-empirical PM3 or Hartree-Fock molecular orbital models, molecular modeling packages for all-atom modeling, and elastic continuum models for structure design. We have built many software components and used virtual reality interfaces which together constitute an implementation of this interactive multiscale simulation platform. Design, simulation, and control of novel bio-nano structures, actuators and robotic systems are developed in Chap. 3 in order to illustrate the proposed interactive multiscale prototyping platform.

The second one integrates a co-prototyping approach based on a dual paradigm, i.e., experimentation and molecular modeling. Usually, nanodevice optimization is only based on experimental investigation of nanodevice characteristics. As all design parameters are not accessible by experimental measurements, the designer is not able to characterize deeply the nanodevice operating principles leading to optimal performances. As alternative, we propose to use jointly modeling tools to support the designer in order to characterize novel phenomenons, evaluate energy densities or new operational principles. The proposed modeling methods uses electrostatic calculation methods with classical molecular dynamics with AIREBO potential. The cooperative work between experiments and simulation paves the way for high-performance nanorobotic components. As illustration of this concept, a co-prototyping methodology applicable to optimized multiwalled carbon nanostructures is proposed in Chap. 4.

References

1. Kitamura, K., Tokunaga, M., Iwane, A.H., Yanada, A.: A single myosin head moves along an actin filament with rectangular steps of 5.3 nm. Nature **397**, 129 (1999)
2. Shingyoji, C., Higuchi, H., Yoshimura, M., Katayama, E., Yanagida, T.: Dynein arms are oscillating force generators. Nature **393**, 711–714 (2001)
3. Montegano, C.D., Bachand, G.D.: Constructing nanomechanicaldevices powered by biomolecular motors. Nanotechnology **10**, 225–331 (1999)
4. Erickson, H.P.: Stretching single protein molecules: titin as a wired spring. Science **276**, 1090–1092 (2003)
5. Mahadevan, L., Matsudaira, P.: Mobility powered by supramolecular, springs and ratchets. Science **288**, 95–99 (2000)
6. Hamdi, M., Sharma, G., Ferreira, A., Mavroidis, D.: Characterization of protein based spring-like elastic joints for biorobotic applications. In: IEEE International Conference on Robotics and Automation, May 15–19, Orlando, FL, pp. 1794–1799 (2006)
7. Ferreira, A., Sharma, G., Mavroidis, D.: New trends in bionanorobotics using virtual reality technologies. In: IEEE International Conference on Robotics and Biomimetics, June 30–July 3, Hong Kong, pp. 89–94 (2005)
8. Requicha, A.: Nanorobots, NEMS nanoassembly. IEEE J. Sens. **91**(11), 1922–1933 (2003)
9. Li, G.Y., Xi, N., et al.: "Videolized" atomic force microscopy for interactive nanomanipulation and nanoassembly. IEEE Trans. Nanotechnol. **4**(5), 605–615 (2004)
10. Shih, C.-Y., Zheng, S., Meng, E., Tai, Y.-C., Liu, Y., Stoddart, J.F.: Nano-to-micro self-assembly using shear flow devices. In: 17th IEEE International Conference on Micro Electro Mechanical Systems, pp. 422–425 (2004)
11. Yin, P.: DNA based self-assembly and nano-device: theory and practice. Doctoral Thesis, UMI Order Number: AAI3181510, Duke University, 2005
12. Humphrey, W., Dalke, A., Schulten, K.: VMD: visual molecular dynamics. J. Mol. Graph. **14**, 33–38 (1996)
13. Taylor, R.M. II: VRPN: virtual reality peripheral network. http://www.cs.unc.edu/Research/vrpn/
14. Stone, J., Gullingstrud, J., Grayon, P., Schulten, K.: A system for interactive molecular dynamics simulation. In: 2001 ACM Symposium on Interactive 3D Graphics, ACM SIGGRAPH, New York, pp. 191–194 (2001)
15. Nelson, M., Humphrey, W., Gursoy, A., Dalke, A., Kale, L., Skeel, R.D., Schulten, K.: NAMD-a parallel object-oriented molecular dynamics program. Int. J. Supercomput. Appl. High Perform. Comput. **10**, 251 (1996)
16. MacKerell, A.D., Bshford, D., Bellott, M., Dunbrack, R.L., Evanseck, J.D., Field, M.J., et al.: All-atom empirical potential for molecular modeling and dynamics studies of proteins. J. Phys. Chem. B **102**, 3586–3616 (1998)
17. Maiti, A.: Multiscale modeling with carbon nanotubes. Microelectron. J. (2007)
18. Liu, Y.J., Chen, X.L.: Continuum models of carbon nanotube-based composites using the boundary element method. J. Bound. Elem. **1**(2), 316–335 (2003)
19. Karakasidis, T.E., Charitidis, C.A.: Multiscale modeling in nanomaterials science. Mater. Sci. Eng. (2006)
20. Ghoniem, N.M., Busso, E.P., Kioussi, N., Huang, H.: Multiscale modelling of nanomechanics and micromechanics: an overview. Philos. Mag. **83**(31–34), 3475–3528 (2003)
21. Feig, M., Karanicolas, J., Brooks, C.L.: MMTSB Tool Set: enhanced sampling and multiscale modeling methods for applications in structural biology. J. Mol. Graph. Model. **22**, 377–395 (2004)
22. Ahmed, A., Gohlke, H.: Multiscale modeling of macromolecular conformational changes combining concepts from rigidity and elastic network theory. Proteins Struct. Funct. Bioinf. **63**, 1038–1051 (2006)
23. Villa, E., Balaeff, A., Schulten, K.: Structural dynamics of the lac repressor-DNA complex revealed by a multiscale simulation. Proc. Natl. Acad. Sci. USA **102**(19), 6783–6788 (2005)

24. Lai-Yen, S.K., Lee, Y.S.: Energy-field optimization and haptic-based molecular docking and assembly search system for computer-aided molecular design (CAMD). In: Symposium on Haptic Interfaces for Virtual Environment and Teleoperator Systems, pp. 233–240 (2006)
25. Pattabiraman, N., Levitt, M., Ferrin, T.E., Langridge, R.: Computer graphics in real-time docking with energy calculation and minimization. J. Comput. Chem. **6**(5), 432–436 (1986)
26. Gao, M., Wilmanns, M.: Steered molecular dynamics study of titin I1 domain unfolding. Biophys. J. **83**, 3435–3445 (2002)
27. Hamdi, M., Sharma, G., Ferreira, A., Mavroidis, D.: Molecular mechanics study on bio-nano robotic components using force-feedback. In: IEEE International Conference on Robotics and Biomimetics, June 30–July 3, Hong Kong, pp. 105–110 (2005)
28. Wollacott, A.M., Merz, K.M.: Haptic applications for molecular structure manipulation. J. Mol. Graph. Model. **25**(6), 801–805 (2007)
29. Daunay, B., Micaelli, A., Régnier, S.: 6 DOF haptic feedback for molecular docking using wave variables. In: IEEE International Conference on Robotics and Automation, May, Roma, Italy (2007)
30. Stewart, J.J.P.: Optimization of parameters for semi-empirical methods IV: extension of MNDO, AM1, and PM3 to more main group elements. J. Mol. Model. **10**, 155–164 (2004)
31. Cramer, C.J.: Essentials of Computational Chemistry, pp. 191–232. Wiley, New York (2002)
32. Kuhn, B., Kollman, P.A.: Binding of a diverse set of ligands to avidin and streptavidin: an accurate quantitative prediction of their relative affinities by a combination of molecular mechanics and continuum solvent models. J. Med. Chem. **43**, 3786–3791 (2000)
33. MacKerell, Jr. A.D., Banavali, N., Foloppe, N.: Development and current status of the CHARMM force field for nucleic acids. Biopolymers **56**, 257–265 (2001)
34. Reference, Monte-Carlo
35. Trombitas, K., Gasser, M., Labeit, S., Jin, J.-P., Kellermayer, M., Helmes, M., Granzier, H.: Titin extensibility in situ: entropic elasticity of permanently folded and permanently unfolded molecular segments. J. Cell Biol. **140**, 853–859 (1998)
36. Zhang, B., Guangzhao, X., Evans, J.S.: A kinetic molecular model of the reversible unfolding and refolding of Titin under force extension. Biophys. J. **77**, 1306–1315 (1999)
37. Isralewitz, B., Baudry, J., Gullingsrud, J., Kosztin, D., Schulten, K.: Steered molecular dynamics investigations of protein function. J. Mol. Graph. Model **19**(12), 13–25 (2001)
38. Hohenberg, P., Kohn, W.: Phys. Rev. B **136**, 864 (1964)
39. Kohn, W., Sham, L.J.: Phys. Rev. A **140**, 1133 (1965)
40. Cramer, C.J.: Essentials of Computational Chemistry. Wiley, New York (2002)
41. Goringe, C.M., Bowler, D.R., Hernández, E.: Rep. Prog. Phys. **60**, 1447 (1997)
42. Rapaport, D.C.: The Art of Molecular Dynamics Simulation. Cambridge University Press, Cambridge (1995)
43. Allen, M.P., Tildesley, D.J.: Computer Simulation of Liquids. Clarendon, Oxford (1987)
44. Tersoff, J.: Phys. Rev. Lett. **56**, 632 (1986)
45. Tersoff, J.: Phys. Rev. B **27**, 6991 (1988)
46. Tersoff, J.: Phys. Rev. B **39**, 5566 (1989)
47. Brenner, D.W.: Phys. Rev. B **42**, 9458 (1990)
48. Stuart, S.J., Tutein, A.B., Harrison, J.A.: J. Chem. Phys. **112**, 6472 (2000)
49. Abell, G.: Empirical chemical pseudopotential theory of molecular and metallic bonding. Phys. Rev. B **31**, 6184 (1985)
50. Brenner, D.W.: The art and science of an analytic potential. Phys. Stat. Sol. (B) **217**, 23 (2000)
51. Robertson, D.H., Brenner, D.W., White, C.T.: On the way to fullerenes: molecular dynamics study of the curling and closure of graphitic ribbons. J. Phys. Chem. **96**, 6133–6135 (1992)
52. Iijima, S., et al.: Structural flexibility of carbon nanotubes. J. Chem. Phys. **104**, 2089 (1996)
53. Huhtala, M., Kuronen, A., Kaski, K.: Carbon nanotubes under bending strain. Mater. Res. Soc. Symp. Proc. **706**, 289–294 (2002)
54. Pettifor, D.G., Oleinik, I.I.: Analytic bond-order potentials beyond Tersoff-Brenner. Phys. Rev. B **59**, 8487–8506 (1999)
55. Li, C.Y., Chou, T.W.: Electrostatic charge distribution on single-walled carbon nanotubes. Appl. Phys. Lett. **89**, 063103 (2006)
56. Hamdi, M., Ferreira, A.: DNA nanorobotics, Microelectron. J. (2008, in press)

Chapter 3
Design and Computational Analysis of Bio-Nanorobotic Structures

1 Introduction

Bio-nanorobotics is a new and rapidly growing interdisciplinary field addressing the assembly, construction and utilization of biomolecular devices based on nanoscale principles and/or dimensions. A key application is for medical target identification in therapeutical diagnosis, medical therapies and minimally invasive surgery (MIS) [1–3]. Modern engineering actuation techniques inspired by nature have been successfully implemented in microrobots evolving in fluidic environments using external electromagnetic fields [4–6]. On contrary, biological approach shows that flagellated propulsion mechanisms of bacteria [7], peptide based nanoGripper [8] or magnetic stereotaxis systems [9] represent a fertile territory for untethered nanoscale machines without the need of external hardware for actuation. Developing nanorobots out of proteins elements requires the merging of two fields of research approaches: the inspiration by nature and biology ("biomimetics") and the inspiration by large scale machines and the traditional machine theory ("machine nanomimetics"). The main goal is to use various biological elements—whose function at the cellular level creates motion, force and energy—as nanometer-scaled machine components. Proteins represent fertile territory for nanoscale machines that produce linear motions in liquid environments. They could act as

- *Molecular springs*: The biological spring proteins are particularly interesting from their elastic, resilient and stability characteristics: stretch and relax without any net energy dissipation, can store and release energy and rectify motion in physical nanomachines in a full reversible way. Spring-like proteins, such as myosin [10] and actin [11] in skeletal muscle, or the contractile stalk of the single-celled peritrish Vorticella Convallaria [12] driven by the spasmoneme organelle.
- *Transmission elements*: Just as muscles magnify forces and movements by using geometrical hierarchy mechanisms, these unusual nanomechanisms uses similar principles: cumulative nanoscale changes in elementary protein subunits are amplified by their linear arrangement in filaments in order to perform micro-scale mechanical work. We can cite amplification nanomechanisms based on molecular hinges actuated by calix[4] arenes proteins [13], switchable molecular com-

M. Hamdi, A. Ferreira, *Design, Modeling and Characterization of Bio-Nanorobotic Systems*,
DOI 10.1007/978-90-481-3180-8_3, © Springer Science+Business Media B.V. 2011

pounds in the shape of bistable [14], rotaxane molecules [15, 16] for nanomechanical amplifiers, or linear ratchets mechanisms [17] similar to macroscopic comb-drive transmissions.
- *Biological motors*: The most familiar motors are composed of elementary muscle myosin formed by protein actins [18], as well as dynein [19] and kinesyn [20] families of proteins, to provide motion of single molecules according to their substantial conformational changes in response to environmental stimuli. At a larger scale, complex proteins are based on electrochemical gradients used by the F1F2-ATP synthase motors [21] or flagellar motors [22] for bacteria to move in three-dimensional liquid space. Today DNA is being used extensively as a construction material for the fabrication of nanoscale systems due to its biological self-assembly intermolecular interactions [23].

By assembling different components together in the proper proportion and orientation, nanodevices with multidegrees of freedom could be generated, able to apply forces and manipulate objects. These long term-objectives of creating protein-based nanodevices necessitates the development of viable design methodologies and computational methods involving basic knowledge on nanoscale kinematics, dynamics and control. The experimental techniques usually employed for protein/molecule characterization have one limitation consisting of the difficulty of direct observation and quantification of protein deformation occurring during mechanical tests. Due to the small scale structure of the proteins or DNAs it is difficult to anchor them, grab them and control their behaviour during the motion experiments. Therefore there is a need to develop alternate computational methods to study the mechanisms by which these bio-nanodevices operate and also to estimate the limitations of the involved mechanochemical conversions.

The objective of this chapter is to present the design hypothesis, architectures and computational results of novel bio-nanorobotic structures. Towards this end, we used the framework presented in Chap. 2 that couple multiscale, multiphysics and virtual reality advanced techniques. The present study demonstrates that the mechanical properties of protein based nanodevices can be controlled by using rational design based on protein and/or DNA engineering principles.

First, protein-like and DNA-like elastic joints elements for structural links have be carefully studied and simulated through Steered Molecular Dynamics (SMD) to understand the functional limits. We simulated the restoring forces involved under various external mechanical stress (stretching, contraction, shearing, bending) to predict the type of force spectra, reversibility, degrees of freedom and irreversible work that may be expected from single-molecule protein manipulation experiments. Based on these computational studies, biological passive elasticity of striated muscle sarcomere nanomechanisms composed of serially linked tandem proteins are analyzed. Multiscale simulation results show clearly the main advantages of the proposed method in terms of computational time, accuracy of simulation and energy optimization. Based on literature experiments, the proposed multiscale design and model is validated.

Second, we propose novel concepts of DNA-based actuation nanomechanisms. As DNA molecule undergoes substantial conformational changes in response to environmental stimuli (temperature, acidic concentration, salt, ionic level), different nanodevices are designed, simulated and analyzed. The first prototype considers an encapsulated DNA molecule acting as nanoscale actuator inside carbon nanotubes in a water solute environment. We report dynamics atomistic-continuum simulations of the dynamic processes towards the prototyping of biological servo nanoactuators (termed, DNA@MWNT). Under the temperature-dependent conformational relaxation of DNA encapsulated in a double-walled carbon nanotube, a controllable and reversible linear motion has been investigated using molecular dynamics simulation. A second prototype describes a new biomolecular gripper. It is composed of a doublestrand DNA (dsDNA) molecule with two single walled carbon nanotubes as nano endeffectors operating in mildly acidic medium (pH = 5.5). Reversible finger's displacement motions are computationally simulated and analyzed during the heating-cooling process.

2 Characterization of Protein-based Nanosprings

Biological springs can be classified as active mechanochemical devices that store the energy of conformation from various bonded and non-bonded interactions as potential energy. Inspired by this concept, we started our studies by the characterization and modeling of different molecular nanosprings with adaptable elasticity and structure which hold promise as passive nanomechanisms: unfolding and refolding of proteins could be a mechanism by which tension is maintained as a protein is extended or relaxed. In order to draw analogies between mechanical protein methodologies and robotics methodologies, several protein-like elastic joints for structural links have be carefully studied and simulated to understand their functional and elastic limits when subjected to external forces (stretching, shearing and bending). To characterize the mechanochemical energy conversion and to study the elasto-mechanical properties of these biological springs we applied steered molecular dynamics (SMD) technique. SMD studies of different protein structures have been conducted (see Fig. 3.1): α-helices (AHs) structures (deca-alanine, Repressor of primer protein, DNA, collagen) and β-sheets (BSs) structures (fibronectin, titin, fibrillin, silk). These molecular structures are derived from X-ray diffraction and NMR data and obtained in the Protein Data Bank (www.rcsb.org/pdb/). It allows to predict the type of force spectra, reversibility, degrees of freedom and irreversible work that may be expected from molecular springs. Standard models for the elasticity of unfolded proteins are based on entropic arguments in combination with enthalpic contributions. For each protein, we identified a force model: linear Hooke model, or exponential models such as inextensible worm-like chain (WLC) model or freely jointed chain (FJC) model.

Methodology of Simulation

All simulations were carried out at a temperature of 300 K (through Langevin damping with a coefficient of 5 ps^{-1}) with temperature rescaling performed every 10

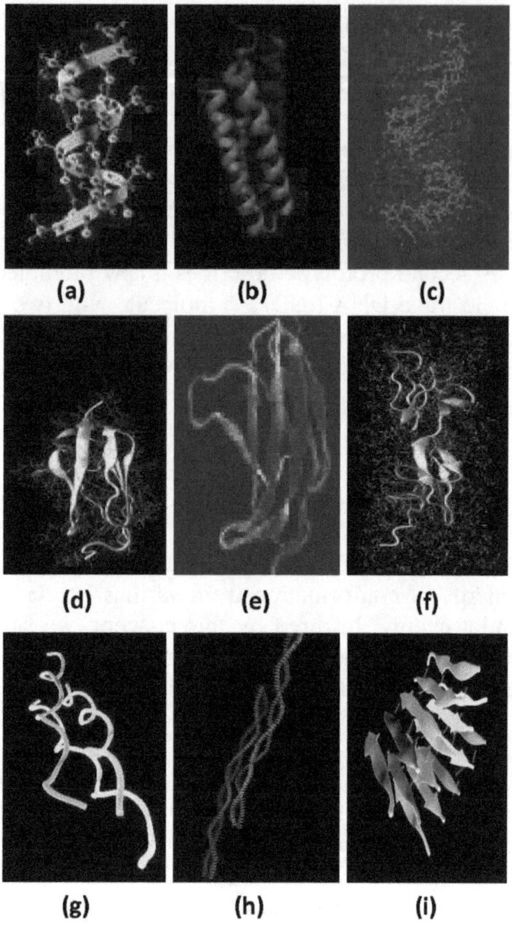

Fig. 3.1 Overview over the nine protein structures considered with loading conditions (stretching, bending, shearing) used to probe the mechanical signature of each spring-like protein. Type 1: α-helices (AHs) structures (**a**) simple-helical deca-alanine, (**b**) four-helical repressor of primer protein and (**c**) double-stranded DNA molecule. Type 2: β-sheets (BHs) structures (**d**) immunoglobubin-like protein I27, (**e**) fibronectin-type III and (**f**) fibrillin protein. Type 3: Self-assembled (SAs) structures (**g**) collagen protein, (**h**) actin protein and (**i**) silk protein

timesteps and a pressure of 1 atm (using the Langevin piston method). In more detail, the protocol is as follows:

1. *Energy minimization*: The first step is the energy minimization of the biomolecular structure in order to remove any strong van der Waals interactions that may exist which might otherwise lead to unstable simulations. At this point, all proteins were solvated by a periodic box of dimensions covering the module by at least five layers of water molecules with the TIP3P model [24]. The water box dimensions were adjusted with respect to each protein structure. Non explicit medium representation, characterized by the dielectric constant ε was considered. The aqueous medium was represented by $\varepsilon = 78$, as proposed by Israelachvilli [25] for water at 27 degree Celsius (300 K). The entire box of water is overlayed onto the protein and those water molecules that overlap the protein are removed. Energy re-minimization is realized by means of the steepest-descent algorithm with a convergence of 10^{-5} kcal/mol.

2. *Heating up process*: Initial velocities at a low temperature are assigned to each atom of the system and Newton's equations of motion are integrated to propagate the system in time. During the heating phase, initial velocities are assigned at a low temperature and the simulation is started periodically, new velocities are assigned at a slightly higher temperature and the simulation is allowed to continue. The entire system gradually was heated up to 300 K in increments of 30 K for time intervals of 1 ps, while leaving the box volume unchanged.
3. *Thermalization*: When the desired temperature is reached, the equilibration procedure consists to run the simulation until that the structure parameters, i.e., pressure, temperature and energy, become stable with respect to time. During equilibration at a temperature of $T = 300$ K, the water molecules composing the box were harmonically restrained to their original positions to maintain the shape of the water bubble. An equilibrium simulation at least 1 ns in length was performed for each protein.
4. *Steered Molecular Dynamics*: The final simulation step was to run Steered Molecular Dynamics (SMD) simulations [25]. Various deformations properties of mechanical proteins can be simulated when subjected to external stretching.

Mechanical Characterization of Protein Based Spring-like Elastic Joints for Nanorobotics Applications

In this section we focus on the comparative analysis of the mechanical signatures of nine basic protein structures. these results were achieved through simulations at time-scales approaching microseconds. The most important result of these studies is the elucidation of the mechanical behavior at experimentally accessible and physiologically relevant deformation rates. In order to reduce the length of the chapter, we decided to report in detail only the mechanical signature of four basic protein structures, e.g., α-helix deca-alanine protein, Immunoglobulin-like molecule and DNA molecule.

1. *α-helix deca-alanine protein* (Fig. 3.1(a)) A canonical right-handed α-helix was built using oligopeptide composed of 10 alanine residues in order to study helix-to-coil transition of α-helix deca-alanine in vacuum and in solvent. In vacuum at room temperature ($T = 300$ K), the stable configuration of deca-alanine is a α-helix. After energy minimization, deca-alanine protein was solvated in a water box of 50 Å length. The system was minimized for 1000 conjugate gradient steps, then the system was minimized and heated from 0 K to 300 K in 6 ps, the system was subsequently equilibrated for 1 ns. Stretching the molecule by an external force can induce its transition to an extended (coiled) form. This helix-to-coil transition represents a simple but basic folding system acting as a molecular spring. The parameter λ is varied from 15 to 35 Å with a constant pulling speed v of 0.1 Å/ps. A force constant of $k = 500$ pN/Å is used in order to allow the end-to-end distance x closely follow the constraint center λ. The

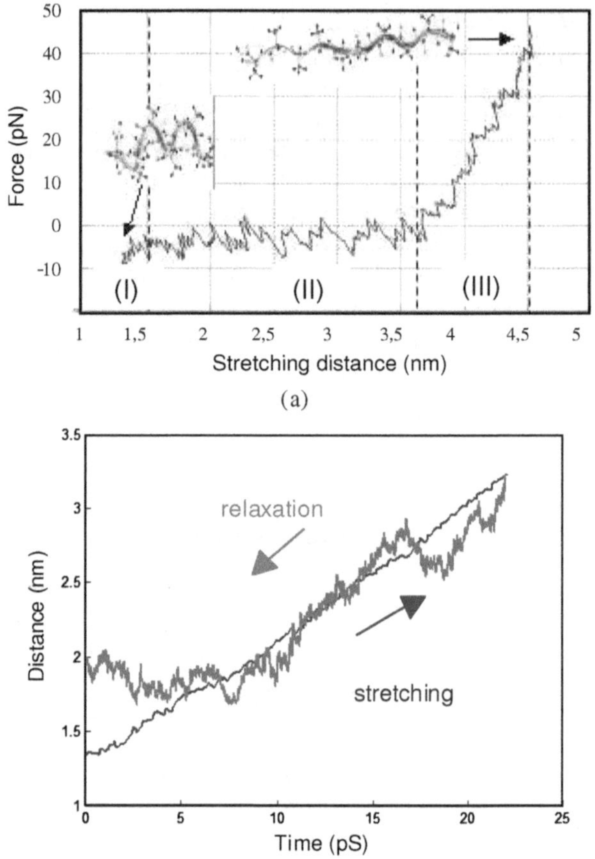

(a)

(b)

Fig. 3.2 α-helix to coil stretching. Force curve and reversibility done by forward pulling (stretching) and backward pulling (contracting) with $v = 1$ Å/ps. For the forward pulling, the position of the constraint center x is varied from 15 to 35 Å; for the backward pulling, from 35 to 15 Å

external work curve is defined as [26]:

$$\langle W \rangle = -kv \int_0^t dt' f(t'). \tag{3.1}$$

Figure 3.2 shows the maximal force (before breaking) as a function of stretching distance. Depending on the sign of v, the external work can be defined for either stretching or contracting motion of the protein. Figure 3.2(a) shows two distinct conformational transitions that provoke the conversion of α-helix to an extended form approaching the coil conformation.

The force increases rather smoothly to almost 10 pN up to the length of the protein. Once this transition is completed, the hydrogen bonds start to break and the structure further extends toward the coil conformation (intrinsic elastic regime), as shown in Part II, of the force curve ($\sim$700 pN). At the lowest forces, the molecule behaves as a Hookian spring and its extension is proportional to the force applied at the end with a reversible motion. A useful approximation

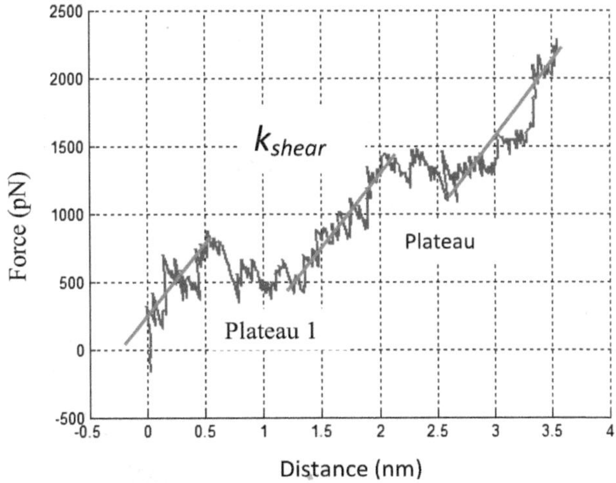

Fig. 3.3 Force-extension behavior of α-helix deca-alanine protein when perturbed by a lateral shearing force F_{shear}

for spring constant k_{stret} is given by the inextensible worm-like chain (WLC) model [27]. The WLC model of entropic elasticity predicts the relationship between the relative extension of a polymer (z/L) and the entropic restoring force (f) through

$$f = \left(\frac{k_B T}{A}\right)\left[\frac{z}{L} + \left(\frac{1}{4(1-\frac{z}{L})^2} - \frac{1}{4}\right)\right] \tag{3.2}$$

where k_B is the Boltzmann constant, T is absolute temperature, A is the persistence length, z is the end-to-end length, and L is the length. Then, if the stretching force increases gradually until it reaches its stretching limit (~3600 pN) with an irreversible motion in Part III. Figure 3.2(b) shows the nonlinear reversibility of the protein when relaxed. It shows clearly that the elastic behavior of the protein can be used as a passive control element. The previous results showed that the native structure is not destroyed under normal physiological conditions. In some unnatural conformations, lateral shearing and bending forces applied on a protein molecule has been simulated. These tests simulate disturbances of the bio-nanorobotic component under various operating conditions. For these conformations, Fig. 3.3 presents the lateral forces obtained. As shown in Fig. 3.3, the lateral Hookian spring k_{shear} has high stiff spring value that is able to counteract microenvironment variations and mechanical disturbances of a bio-nanorobotic platform. In this case, in contrast to stretching described above, the force variation is roughly monotonic with different plateaus leading to different Hookian spring values. The extension of the Hookian spring is proportional to the force applied at its end such as:

$$f = \left(\frac{k_B T}{A}\right)\left(\frac{x}{L}\right). \tag{3.3}$$

This is explained by a successive rather than a concurrent rupture of the hydrogen bonds joining the strands. Shearing is largely limited to breaking hydrogen

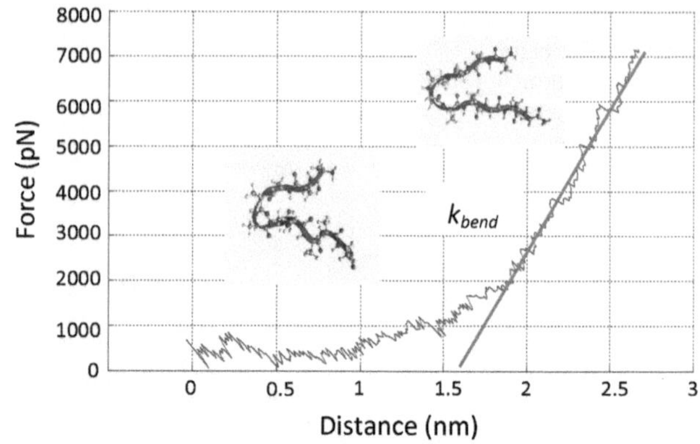

Fig. 3.4 Force-extension behavior of α-helix deca-alanine protein when perturbed by a bending force F_{bend}

Fig. 3.5 Structure of water-DNA system ionized with 3O Na^+ ions. *A*, *C*, *G* and *T* stand for adenine, cytosine, guanine and thymine

bonds as there is little conformational change in the extended peptide backbones. Conversely, Fig. 3.3 shows a linear variation of force behavior when considering pure bending deformation of the protein. It shows that for little length deformation, it requires low constant force. After a threshold value, lateral bending must break the van der Waals interactions of the entire surface, leading to high forces and strong length dependence. Finally, these results indicate that it may be possible to obtain uncoupled mechanical spring behavior of the protein: stiff in lateral and compliant in longitudinal direction.

2. *Double stranded DNA (dsDNA) molecule* (Fig. 3.1(d)) Chemically, DNA is a long polymer made up of a linear series of subunits known as nucleotides. Structurally, DNA is usually found as a double helix, with two strands wrapped around one another. However, DNA can adopt other configurations and can also exist in single-stranded forms. Double-stranded deoxyribonucleic acid (dsDNA) has sparked the renewed interest in the force versus extension of polymers for biomolecular springs. In the following simulations we choose a B-DNA molecule solvated in water with 30 Na^+ ions added to neutralize the charge (see Fig. 3.5). The water-DNA system was gradually heated over 7 ps to 300 K, and then equilibrated with a thermal bath at 300 K for another 7 ps.

   The dynamics of the DNA as shown in Fig. 3.6 was performed with the double-stranded terminations stretched and the other end fixed. This elastic behavior is thus purely entropic. For very low tension $f \leq 1$ pN, the restoring force is provided by "entropic elasticity". In the absence of any force applied to its ends, the DNA's RMS end-to-end distance (chain length, L) is small compared to its contour length defined as the maximum end-to-end distance (maximum length, L_0) and the chain enjoys a large degree of conformational disorder.

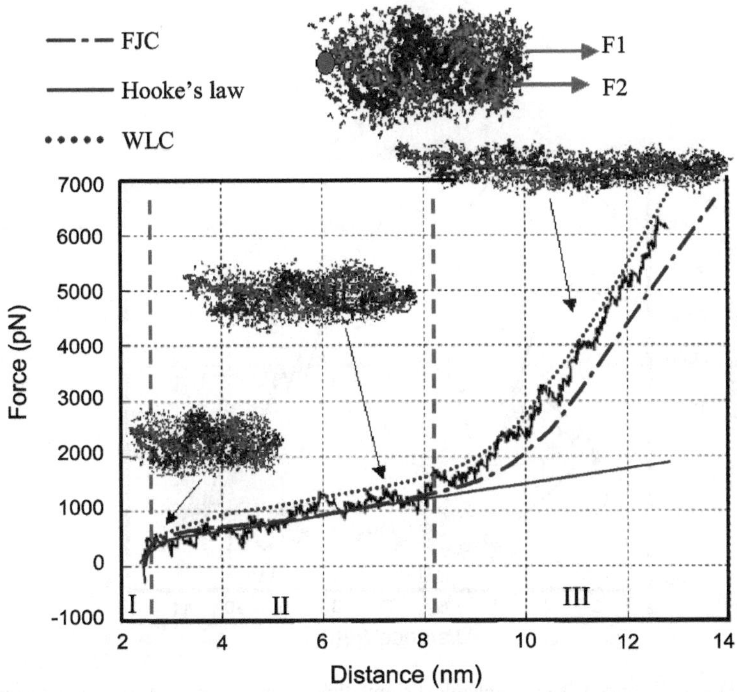

Fig. 3.6 Double-stranded DNA mechanical simulations: Force-extension curve when applying a stretching force

Stretching DNA reduces its entropy and increases the free energy. The corresponding force f increases linearly as a Hooke's law with the extension L:

$$f = \left(\frac{k_B T}{A}\right)\left(\frac{x}{L}\right), \quad L \ll L_0. \tag{3.4}$$

The length A is known as the "thermal persistence length" of DNA and is of the order 50 nms. For higher forces ($f \geq 10$ pN), the end-to-end distance L is close to L_0 and the elastic restoring force is due to distortion of the internal structure of DNA. In this regime, the force extension curve can be approximated by two models that are often used to describe the entropic elasticity of DNA. In the freely jointed chain (FJC) model, the molecule is made up of rigid, orientationally independent Kuhn segments whose length, b, is a measure of chain stiffness. The resulting entropic elastic behavior can be summarized in the force-extension relation:

$$\left(\frac{Z}{L_{tot}}\right) = \coth\left(\frac{fb}{k_B T}\right) - \frac{K_B T}{fb} \tag{3.5}$$

defining the well-known Langevin function. Expanding (3.5) gives the effective spring constant for the FJC as $k = \frac{3K_B T}{b}$. The terms L_{tot} represent the total length

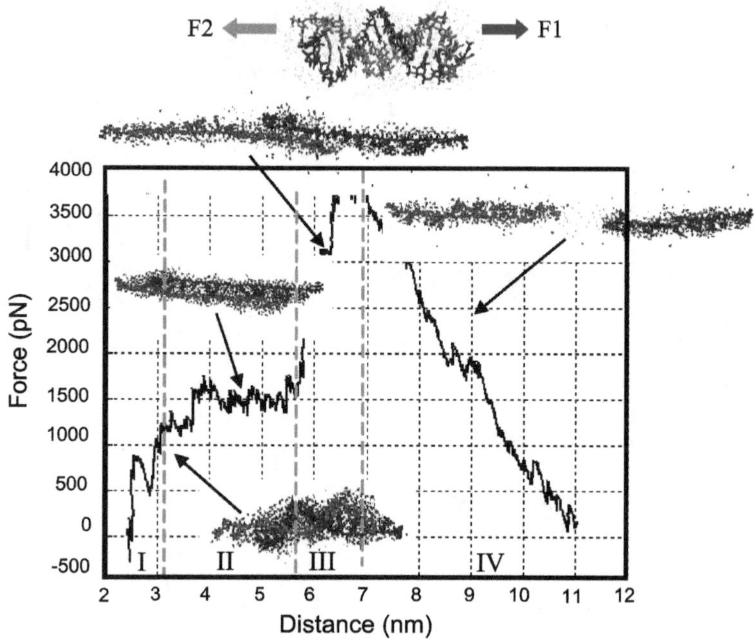

Fig. 3.7 Double-stranded DNA mechanical simulations: Force-extension curve when applying an anti-parallel stretching

of the protein and f the stretching force. The alignment of segments by tension is described by Boltzmann distribution. In the inextensible worm-like chain model (WLC) model [27], the molecule is treated as a flexible rod of length L that curves smoothly as a result of thermal fluctuation. The WLC model of entropic elasticity predicts the relationship between the relative extension of a polymer $(\frac{z}{L})$ and the entropic restoring force (f) through

$$f = \left(\frac{k_B T}{A}\right)\left[\frac{z}{L} + \left(\frac{1}{4(1-\frac{z}{L})^2} - \frac{1}{4}\right)\right] \quad (3.6)$$

where k_B is the Boltzmann constant, T is absolute temperature, A is the persistence length, z is the end-to-end length, and L is the length. Results, shown in Fig. 3.7, indicate that, even though the FJC model can describe the behavior of dsDNA in the limit of low and intermediate forces, it fails at high forces. The data are fit to a WLC model solved numerically assuming $A = 43$ nm. The FJC curve assumes $b = 2P = 106$ nm. Both models behave as a Hooke's law for low stretching forces. In the opposite limit of large force the end-to-end distance approaches molecular contour length and subsequently the force-extension curve begins to rise quickly. At these large forces, DNA can be extended slightly beyond its contour length, which is accounted for by the elastic stretching modulus l. Experiments performed with laser tweezers [27], magnetic traps [28], and

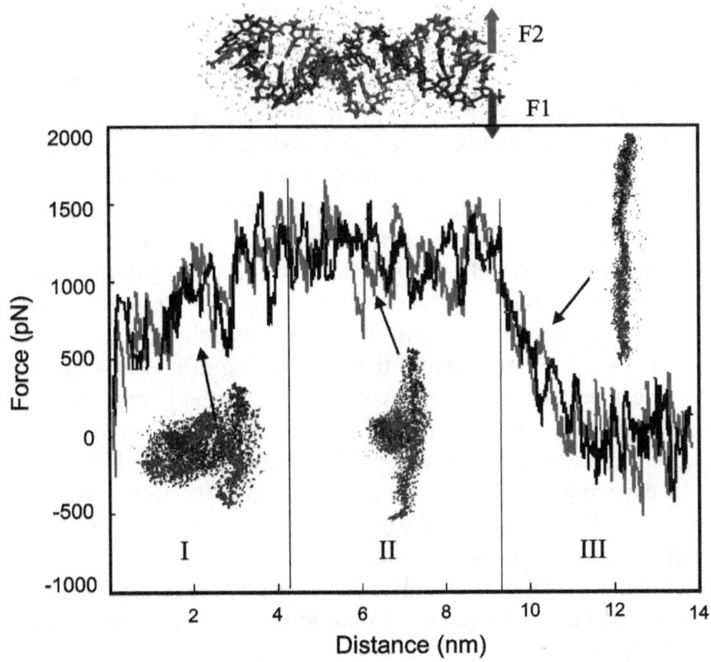

Fig. 3.8 Double-stranded DNA mechanical simulations: Force-extension curve when applying an anti-parallel shearing

AFM-based single molecule force spectroscopy [29] describe accurately the elastic behavior of DNA at forces up to 10 pN. For higher forces, this experimental behavior presents a sharp structural transition leading to a strong overstretching of the molecule that cannot been reflected by the WLC model. The molecule suddenly changes from its canonical B-form contour length to a so-called S-form DNA (B-to-S overstretching transition). In the case of anti-parallel stretching (Fig. 3.8), the force-extension curve is quite similar for the Parts I and II when the dsDNA is unfolded. When it is totally stretched, lateral mechanical unzipping of the hairpins occurs sequentially leading to an increase of the stretching force (Part III) until to reach its point of rupture (Part IV). These simulation results are consistent with the experimental ones given in [30]. The multiple rupture events between different strands occurring at different times during the separation (Part II) are clearly measured with the last rupture event (Part III) before separation (Part IV). However, it should be noticed that the AFM-based force spectroscopy experiments show differences in the size of force curves. As the AFM can measure forces ranging from picoNewtons to hundreds of nanoNewtons, the authors in [31] investigated mainly the forces holding the DNA together (H-bonding and base stacking) than the backbone conformation rupture forces.

When considering anti-parallel shearing of dsDNA (Fig. 3.8), a mechanical longitudinal unzipping of the 12 hairpin DNA occurs sequentially (Part I) after a phase of extension of dsDNA structure (Part II) before the rupture point (Part III).

As it is shown, the strand unzipping occurred abruptly at 500 pN and displayed a reproductible "sawtooth" force variation pattern with an amplitude of ±10 pN along the DNA. We found that the force level necessary to trigger the opening (maximum amplitude of a given sawtooth) along the sequence increases with the GC content of the peak that blocks the progression. These triggering events were observed here in the 40∼60 percent range; a linear fit in that domain of the force versus GC gave a rough estimate of ±15 pN for opening a 100 percent GC sequence and 10 pN for a 100 percent AT sequence. Validation of this molecular stick-slip has been analyzed experimentally in [32, 33]. Simulations shows that releasing the opening force (not shown), the two single strands reannealed, and a new stretching cycle could be engaged. In [32], the authors reported a direct measurement on the forces involved in the unzippering of a long segment of DNA. However, the mechanical forces involved in the strand separation were lower than our simulations (in the range of 10–15 pN). This difference was related to the sequence and the elastic nature of the mechanical B-DNA configuration used in the experiments. DNA, in particular, as recently been subject to thorough experimental and theoretical investigations in many groups, with a view to understanding the elastic properties of DNA and its stability against force-induced overstretching [34, 35]. These studies all focus on how double-stranded B-DNA is stretched and bent with respect to variation of medium environment. Considering the interesting stretching characteristics of dsDNA (see Fig. 3.6), we designed a passive DNA-based elastic joint composed of two single walled carbon nanotubes (SWNT) as structural components (see Fig. 3.9(a)). The covalent linkage between DNA-SWNT are realized at the terminations through atoms of carbon Fig. 3.9(b). As shown in Fig. 3.10, the passive dsDNA-based joint is fully reversible in the domain of linear elasticity of the DNA molecule (Part I and II). The spring constant is estimated to be as $K_{DNA} = 780$ N/m with a thermal bath at 300 K and constant acidic pH = 5.

3. *Immunoglobulin-like domain* (*Ig-type*) (Fig. 3.1(d)) β-sandwich structures are found in a variety of mechanicalproteins composed of multiple, individually folded protein domains. The forced folding and unfolding of proteins using SMD simulations have been mostly reported for two important examples: the immunoglobulin (Ig)-type fold [36, 37] and the fibronectin-type fold (the most common of which is fibronectin type 3 or FN-III) [38]. Force-induced extension of the protein titin, for example, which is responsible for the passive elasticity of muscle, can cause its constituent Ig and FN-III domains to unravel. Currently,

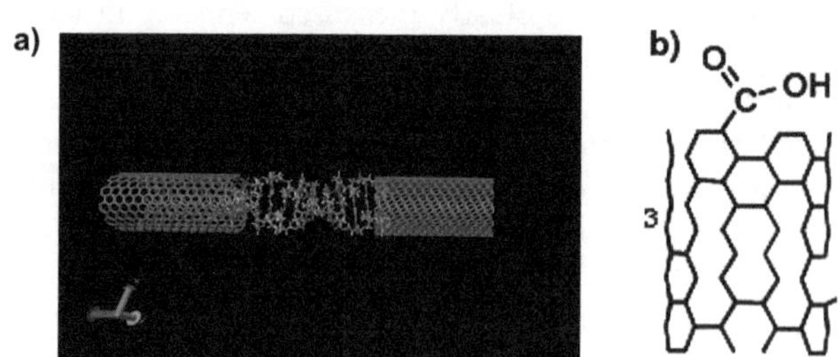

Fig. 3.9 (**a**) Passive joint composed of a dsDNA protein attached at both ends by carbon nanotubes (CNT). The water molecules are not shown. (**b**) Schematic illustration of chemical scheme for carbon nanotube-DNA linkage

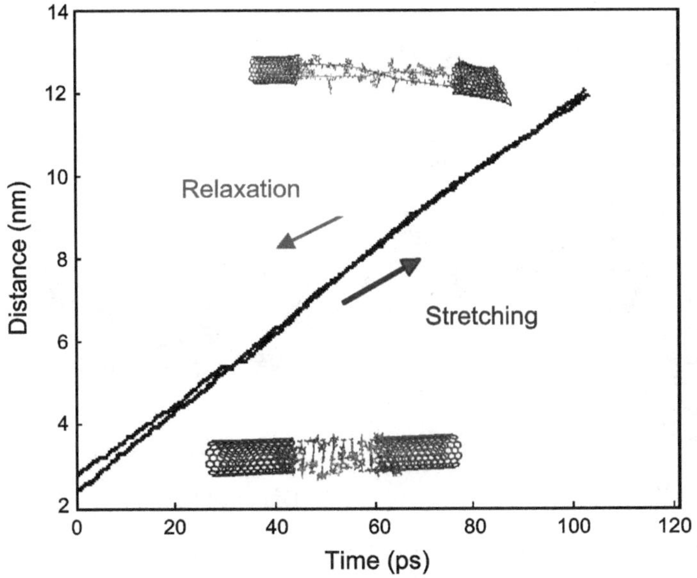

Fig. 3.10 Reversibility of the DNA-based elastic joint during a stretching-relaxation phase

I27 is the only I-band Ig with an experimentally solved structure and hence has been selected for investigation (PDB; entry code 1TIT). The Titin I27 domain was placed in the center of the water box of 70 Å length and equilibrated with a thermal bath of $T = 300$ K. First, the system was minimized for 2000 conjugate gradient steps. Following the minimization, the system was heated from 0 K to 300 K in 10 ps and was coupled to a 300 K heat bath for additional 10 ps. The temperature control was released, and the whole system was subsequently equilibrated for 1 ns. The SMD simulations were carried out by fixing the C-atom of the N-terminus of I27 and applying external forces to the C atom of the C-terminus. The simulation began with an equilibrated folded structure and was stopped when a fully extended polypeptide was obtained. The extension of I27 was performed with a pulling speed $v = 0.5$ Å/ns and was stopped when the extension reached 33 nm. The extension domain (Fig. 3.11) is divided into four sections: I. preburst at extension of 4 nm during which the protein maintains β-sheet structure and the external force remains smaller than 1500 pN; II. major burst immediately after the preburst burst at extension of 8 nm; III. post-burst at extension of 27 nm during which the protein unravels; IV. pulling of fully extended chain up-to an extension of 33 nm. The simulated force extension profile is in accordance with [27, 39].

Other simulations showed similar features of the unfolding process and force profiles with only small variations in force peak value and degree of extension at the force peak. Results of stretching-relaxation curve (Fig. 3.12) show a good reversibility of the protein motion when completely relaxed. Its behavior during extension might be modeled as series of elastic spring with a viscous element

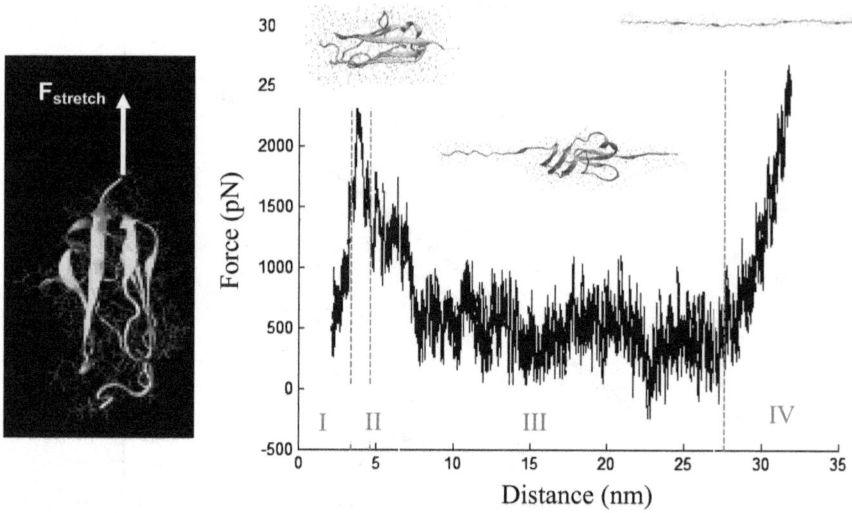

Fig. 3.11 Titin I27-Ig module stretching simulations. Force-distance curve

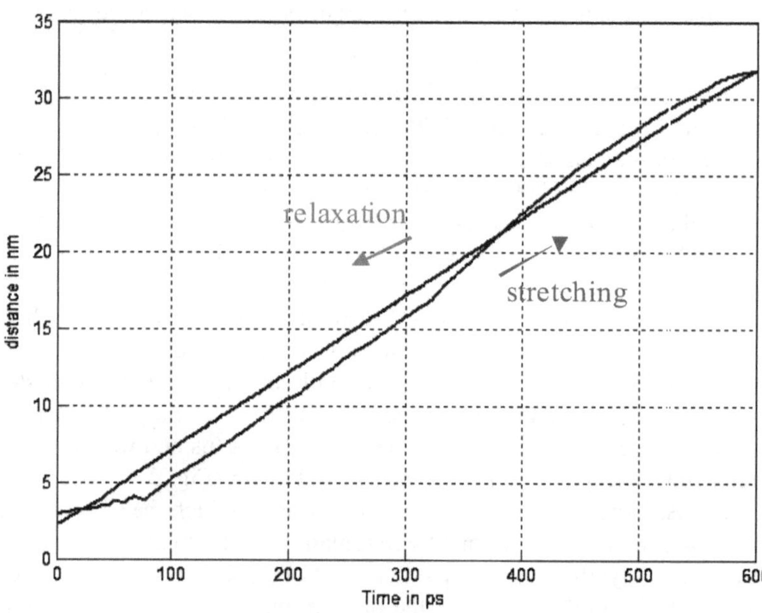

Fig. 3.12 Titin I27-Ig module stretching simulations. Stretching-relaxation curve for reversibility

corresponding to the unfolding of the individual I27 domain. Stretch would result first in straightening of the Ig domain chain (corresponding to the preburst-part I) as an entropic spring. The tightly folded Ig domain might function as a "shock absorber" (parts II: major burst and III: post-burst) by reversible unfolding only

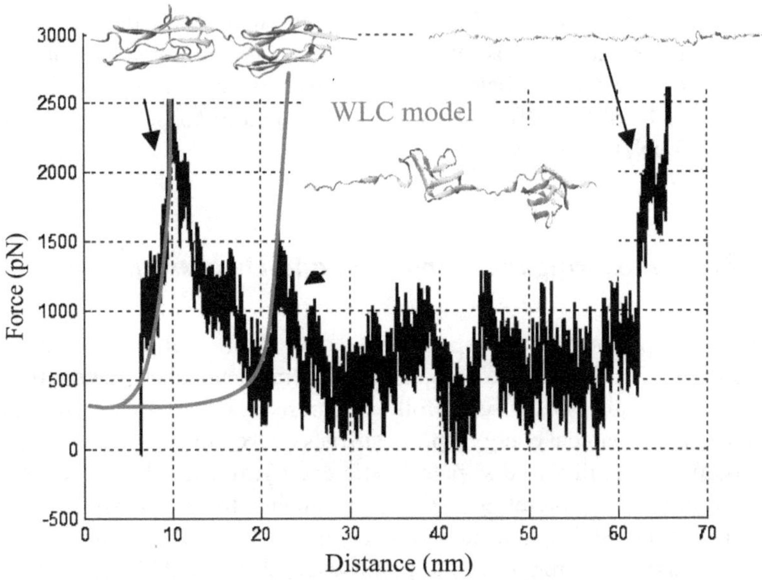

Fig. 3.13 I27-I1 stretching simulations. Force-distance curve of molecular spring Ig modules connected in series

in the case of extremely high stretching forces. This structure allows avoiding the complete rupture of the protein due to overstretching.

As example of serial Ig multi-repeat, Fig. 3.13 shows the typical force-extension curve by stretching two immunoglobulin-like proteins (I1 and I27-Ig modules). The simulation parameters are $v = 10$ Å/ns, $T = 300$ K and a force constant $k = 500$ pN/Å. The initial part of the force-extension curve is fitted with the WLC model to obtain the entropic spring of I27-Ig module. After the unfolding of the I27-Ig module, the second I1-Ig module unfolds and can be also fitted by a WLC model. The extension of (I27-I1) domains can be represented by an elastic spring. When considering several Ig modules connected in series, the force-distance curve behaves as a sequential unfolding of the Ig modules. The number of force peaks arising in the profiles is equal to the number of Ig domains involved in the stretched protein. Its behavior during extension might be modeled as series of elastic springs with different stiffness together with the viscous elements corresponding to the unfolding of individual domains. These simulation results are in accordance with the mechanical properties of human myomesin domain experimented in [40]. Stretching of homomeric polyproteins, constructed of Ig and FN-III domains of human myomesin, produces a typical saw-tooth pattern in the force-extension curve which function as a passive reversible spring with adaptable elasticity.

As shown in Fig. 3.13, titin generates a restoring force based on the mechanism of entropic elasticity. The elasticity of entropic components allows titin to be extended fully reversibly at physiological forces, without the need to unfold the Ig domains as shown in [41, 42]. The elastic component arises from the extension of the serially linked Ig domain chain. Subsequent re-extension of the protein repeat demonstrates a recovery of folded domains that is dependent on the time interval between consecutive extensions. In addition, lateral shearing and bending stiffness have been evaluated to $k_{shear} = 280$ pN/nm and $k_{bend} = 925$ pN/nm, respectively.

Molecular Nanosprings: Comparison and Discussion

In this section, a performance comparison and classification of the nine proteins are presented in Fig. 3.1. In order to compare the key mechanical characteristics of the basic protein structures, we chose the following traits: elasticity (that is, the ability to store and reuse mechanical energy: spring stiffness, reversibility, Young's modulus), energy dissipation (ability to dissipate elastic energy: relaxation time, energy storage, reversibility error), tolerating large displacements (that is, the ability to tolerate large displacements: strain) and fracture strength due to rupture of H-bonds clusters (that is, the maximum strength at the point when all H-bond clusters are broken: fracture strength). A quantitative comparison of the elasto-mechanical properties of protein-based joints are reported in Table 3.1. It is clear that we have treated only idealized cases in this modeling study and that we are using a very simple environmental representation. However, the present study shows clearly some important characteristics of molecular springs. All of them exhibit a conformational bistability of latched and unlatched states with different elastic features. In the case of α-helix structures (deca-alanine, ROP, fibrillin) the protein backbone features an initial linear elastic regime followed by a large plateau regime at which the helix "slowly" uncoils, releasing turn at an time (simulatenous breakage of H-bonds). This plateau regime is characterized by an almost flat force-extension behavior before by a steep increase in the force-extension relation (linear Hookian regime) once the backbone covalent bonds upon complete uncoiling of the α-helix structure. Deca-alanine and fibrillin behave as elastic Hookian springs in a wide range of extension but their main drawback are their poor length change ratio ($\Delta l/l$) and relaxation time (τ). These proteins are very flexible and cannot generate large forces. The ROP protein possess also an interesting stiff parallel structure well adapted to "large force-small displacement" spring-like elastic joint. In the case of β-sheet structures (fibronectin, immunoglobulin-like protein I27, silk), the rupture behavior show a quite different behavior, featuring a characteristic "soft" nonlinear elastic behavior modeled by a WLC or FJC models. Fibronectin Fn-III10 and Immunoglobulin I27 proteins will be preferred due to their powerful elastic force capable to form serial and/or parallel links for the restoring force of biorobotic components. Their small relaxation time allows to magnify small conformation changes and growth into large and rapid movements. They can be used as linear springs for small extensions (less than 10 nm) and as "shock absorbers" during the sequential unfolding of the proteins. Due to their high length change ratio (from 5 to 10), Fn-III10 and I27 proteins are very flexible in their unfolded state. The self-assembled molecules (tropomyosin, tropocollagen) show continuous stiffening in their force-extension characteristics. In the case of tropocollagen molecule, the characteristic triple helical structure provides the structural basis for the continuous "stiffening" behavior (the modulus increases continuously from 0.7 GPa to 16 GPa at large deformation). The progressive stiffening and relatively large force levels provide significant elastic energy storage and recovery. o sudden, catastrophic failure is observed. The analysis of Table 3.1 points out the good mechanical properties of the proteins which allows the design of stable and stiff molecular structures. When stressed in an unnatural folding pathway, fibrillin, fibronectin, immunoglobulin, collagen and ROP proteins act as one degree of freedom stiff passive joints. Double-Stranded DNA (dsDNA) molecule has interesting characteristics due to their elastic behavior and structural conformation. While traditional simple proteins do not exhibit any appreciable other degrees of freedom, complicated ones like DNA do (3d.o.f: stretching, shearing and twist reversibility). They can be used to construct molecular building blocks by virtue of their conformation pathways. Branched DNA molecules with sticky ends are promising for assembling macromolecular periodic structures.

3 Multiscale Design and Modeling of Protein-based Nanomechanisms

Problem Formulation

New nanostructures using synthetic strategies have recently emerged. Inspired by biological systems, concepts of self-organization and self-assembly of building blocks allows the self-association (in series or in parallel) of particular patterns to form higher order organized biological nanomechanisms. Elementary biological structural components composed of protein-based nanosprings are of prime importance in the elasticity and stability of main biological functions at macroscale, i.e., tendons, muscles, lung and skin. Bioinspired nanomechanisms involved in nature are interesting for the design of protein-based nanorobotic structures. Figure 3.14 shows the giant multidomain fibrillar protein titin composing the sarcomeres of human striated muscles which behaves the strong passive elasticity of muscle. Titin, a 1-μm-long protein found in striated muscle myofibrils, possesses unique elastic and extensibility properties due to the serial assembly of coiled titin molecules. Titin is composed of ~300 repeats of two types of domains, fibronectin type III-like (Fn-III) domains and immunoglobulin (Ig) domains. The Ig domain realizing the connection between the end of the thick filament and the Z-line constitutes the titin-based passive elastic nanomechanism. Figure 3.14(c) shows the molecular kinematic chain modeling the extensible part of the titin molecule used in this study: proximal tandem-Ig, PEVK unique sequence and distal tandem-Ig. The biological inspired nanomechanism involves different phases of elasticity (Fig. 3.14(b)). The I-band part of titin is coiled up when a sarcomere is at its back slack length due to the inherent flexibility of the molecule (i.e., the length to which relaxed muscle

Table 3.1 Quantitative comparison of mechanical properties of elastic proteins, including strain, dominant energy, spring stiffness when considering stretching, bending and shearing loadings, reversibility, relaxation time, energy variation, small-strain elastic modulus, and elastic model

	Strain $\Delta l/l$	Spring stiffness			Reversibility %	Relaxation time τ fs	Energy variation ΔE Kcal/mol	Young's modulus[a] E 10^9 Pa	Elasticity model[b]
		k_{stret} pN/nm	k_{bend} pN/nm	k_{shear} pN/nm					
Deca-alanine	1.305	~440	~32	~710	~30	400	975	0.446	Hooke
Fibrillin	4.17	~40	~250	~100	~10	100	1835	0.612	Hooke
Fibronectin	8.483	~850	~360	~60	~5	160	6355	0.2	WLC
Titin I27	5.769	~875	~925	~280	~5	110	4778	0.61	WLC
Tropocollagen	1.707	~90	–	–	~24	22	1211	16	WLC
ROP	1.2	~40	~190	~240	~5	64	705	2.31	Hooke
DNA	1.476	~200	~450	~220	<5	75	3018	1	FJC or WLC
Tropomyosin	1.625	~470	~20	~100	~5	4.4	3018	1.34	FJC or WLC
Silk	4.8	~60	~50	~52	~5	~64	1790	0.315	WLC

[a]The modulus for protein structures is obtained by normalization of the force by the cross-sectional area of the protein domain, for details we refer the reader to the literature

[b]The identified models are Hookian, Worm-Like Chain (WLC) model or freely jointed chain (FJC) model under large-strain displacements

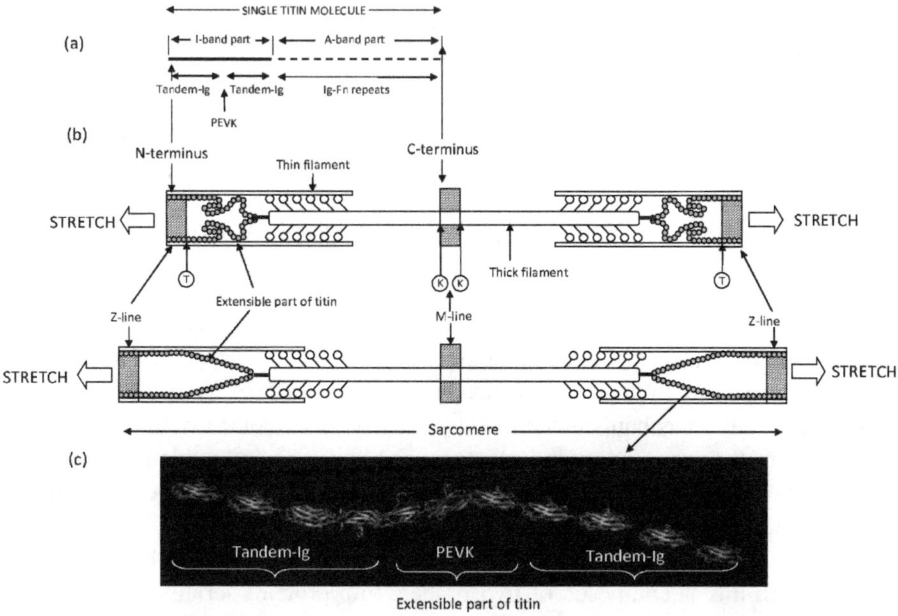

Fig. 3.14 Molecular mechanism of sarcomere passive contraction/relaxation. Cardiac titin's extensible I-band region acts as a biological inspired spring nanomechanism. (**a**) Regions of the titin molecule composed of I-band part and A-band part connected between the N- and C-termini. (**b**) Titin configuration at different sarcomere lengths. Upon sarcomere stretch, initially extension of tandem Ig (immunoglobulin-like) segments dominates followed by dominating PEVK unique sequence extension rich in proline (P), glutanate (E), voline (V) and adlysine (K). Stretching force is provided by elementary myosin motor molecules "walking" on actin filaments. (**c**) Molecular model of Titin-PEVK strand formed by a polypeptide chains

freely shortens). The first elasticity phase only involves straightening of the tandem Ig segments. After straightening, further extension involves the unfolding of the titin polypeptide, giving rise to restoring force. Finally, extension of PEVK unique sequence dominates the stiffness of sarcomere. The magnitude of the force required to stretch titin must be put in the context of the forces operating in cells. The force generated by two to four motor molecules operating in parallel stretch titin completely. Stretching force is provided by elementary myosin motor molecules "walking" on actin filaments. It highlights that the sarcomere behaves as a versatile and adjustable molecular spring with a range of important functions in passive and restoring forces in kinematic nanostructures. Just as muscles magnify force and movements by using hierarchy mechanisms, bio-inspired nanomechanisms using similar principles could lead to the design of new class of bio-inspired nanorobotics. These nano-scale fibrillar nanostructures have equivalencies with macro-scale robotic kinematic chains. The serial and/or parallel composition synthesis of protein-like springs allows to mimic the passive elasticity of biomechanisms found in nature for application in new designs of molecular robotic kinematic structures. The main challenges of the present work focuses on the theoretical understanding, modeling and simulation of long chain tandem-Ig repeats (Fig. 3.14(c)) which implies the development of multiscale and multiphysics strategies from the quantum mechanics to continuum modeling.

Multiscale Modeling and Design of Titin-based Nanomechanisms

The long kinematic polypeptide chain composing the I-band part (see Fig. 3.14) possess extensible and inextensible tandem-Ig repeats. In a simplest way, the sarcomere's contraction/relaxation mechanism can be modeled by a guided hyper-redundant parallel mechanism actuated by multiple molecular myosin motors. As illustration of the proposed multiscale nanorobotic design and modeling approach, we simplified the kinematic study to a 1-dof parallel protein-based robotic platform as shown in Fig. 3.15. Its nanomechanical architecture is composed of three passive parallel links based on elastic and reversible proteins for the extensible tandem-Ig repeats. Each link is composed of two β-sheet titin proteins serially linked. In order to simulate the upward linear motion of the platform generated by the myosin motor, three controllable stretching forces are applied to the upper platform [43]. The three β-sheet titin links acting as spring elements or as a restitution force will bring the platform back to its original position when the myosin motors are at rest. The attachment of the different components is constituted by three carbon atoms at the proteins/graphite interface. The dimensions for the triangular graphite sheet are 5.0 nm by 5.6 nm and the total length of the two proteins is 8.5 nm. In order to model and simulate the stretching/refolding serially linked titin Ig domain, we proposed a solution using several physics for particular length- and time-scales.

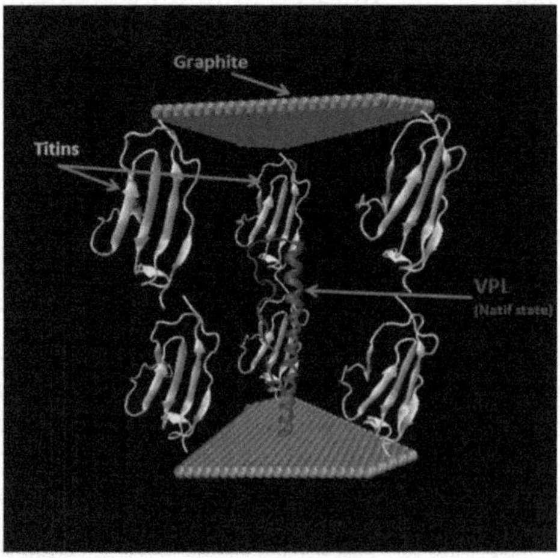

Fig. 3.15 Parallel nanorobotic platform composed of two graphite platforms and three serially linked passive bio-springs to join the two platforms. The upward and backward motions are controlled by linear myosin actuators

Quantum Mechanics (QM)

A wide variety of different procedures or "models" have been developed to calculate molecular structure and energetics among other properties. Different quantum mechanics models are currently developed in the literature, i.e., Hartree-Fock molecular orbital models [44], semi-empirical molecular orbital models [45] and molecular mechanics models [46]. However, these models fail when identifying conformational minima and determining the geometries (bond lengths, angles and charges) of these minima. We developed a specific PM3 semi-empirical model when attaching leucine termini to graphite layer as shown in Fig. 3.16(a). The first task is to determine exactly which parameters are already known, and which will need to be developed from scratch. The unparameterized part of the new residue is the connection between the azote and the carbon. The connection between the azote of LEU and the carbonyl carbon of the ARM residue is called a thioester bond. Leucine-ARM linkage is defined as a new residue defined as a new residue in the CHARMM27 topology file with the name LARM (please refer to Chap 2).

Molecular Dynamics Characterization

As shown previously, the structure of the elastic serial link consists of two domains of globular proteins. The titin protein (I27) is referenced as 1TIT in the Brookhaven Protein Data Bank as entry code "1tit". The domains was solvated in a water box of 70 Å length. First, the system was minimized for 2000 conjugate gradient steps. Following the minimization, the system was heated from 0 K to 300 K in 10 ps and was coupled to a 300 K heat bath for additional 10 ps. The temperature control was released, and the whole system was subsequently equilibrated for 1 ns. Finally, steered molecular simulations (SMD) simulations were carried out by fixing the C-atom of the N-terminus of I27 and applying external forces to the C-atom of the C-terminus. The forces were directed along the vector from the pulled atom to the

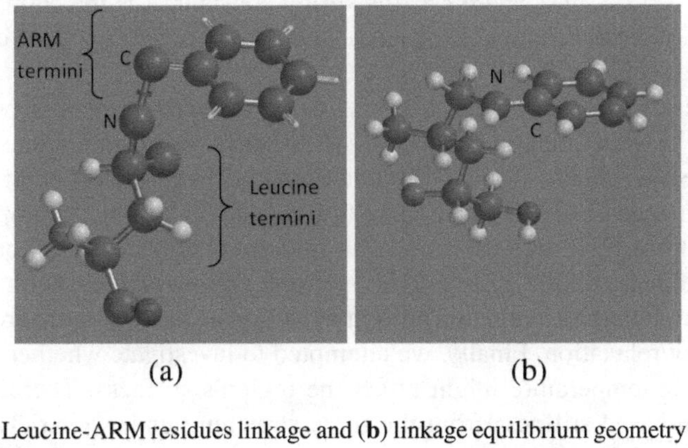

Fig. 3.16 (**a**) Leucine-ARM residues linkage and (**b**) linkage equilibrium geometry at ground state with Semi-Empirical PM3

Fig. 3.17 Steered molecular dynamics using molecular dynamics simulation (QM-MD)

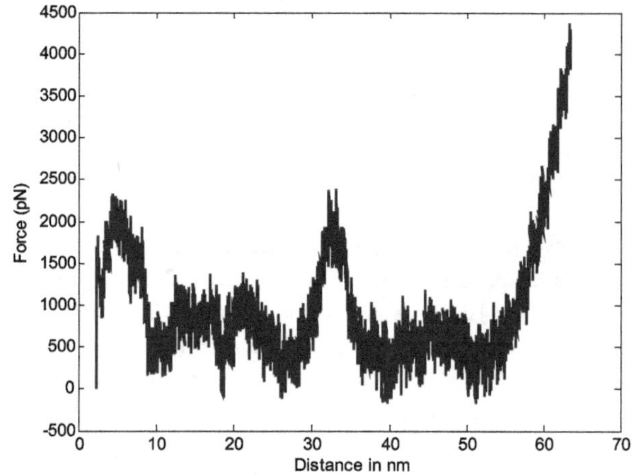

Fig. 3.18 Force-extension curve when applying a stretching force

fixed atom. Constant velocity protocol is used for the SMD simulations with pulling speed 0.5 Å/ps. In the latter case the pulling atom is harmonically constrained with a force $F = -k(x - vt)$, where k is the spring constant, x is the coordinate of the pulling atom, v is the velocity of the atom, and t is the time. The value of the spring constant k was set to 500 pN/nm.

Figure 3.18 shows the typical force-extension curve by stretching two immunoglobulin-like proteins. In comparison to the previous SMD simulations of titin unfolding made by the authors in [47, 48], our results differ from different viewpoints. First, we performed SMD simulation of titin repeats unfolding with optimized energy linkers through PM3 semi-empirical models. Then, stretching forces applied to the parallel nanorobotic structure are in direct correspondence with those experienced in real biomolecular actin-kinesin actuation at the origin of sarcomere passive stretching/relaxation. Finally, we attempted to investigate whether pH changes and/or variable temperature might affect the region's stiffness. Therefore, we determined titin-based stiffness from the force stretching under a variety of pH and temperature conditions.

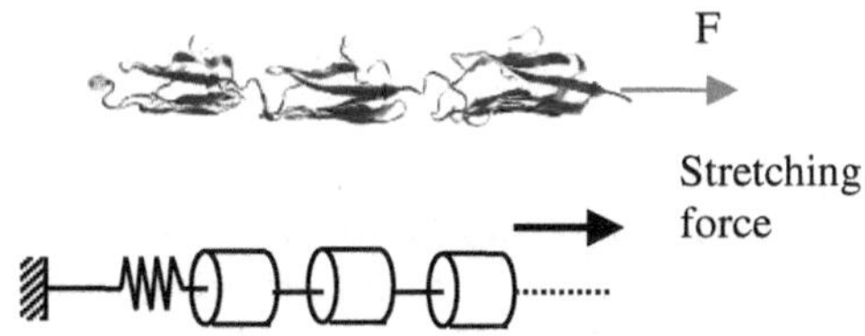

Fig. 3.19 Nanomechanical model of the serially linked titin molecular spring composed of three I27 modules connected in series

Combined Molecular Dynamics and Continuum Approach

The main goal of molecular dynamics studies is to understand the relation between molecular behavior and nanorobot response. Molecular dynamics computing time may take a very long time, thus evaluation of the nanorobotic platform design cannot be achieved through realtime simulation. Combination of molecular dynamics and continuum approaches is investigated. The analysis of Fig. 3.18 shows that the initial part of the force-extension curve is fitted with the worm-like chain model to obtain the entropic spring of "1tit" serial module. After unfolding, the second module unfolds and can be also fitted by a WLC model [49]. The extension of "1tit" serial domains can be represented by an elastic spring. Based on these comments, Fig. 3.19 suggests a nanomechanical portrait model: its behavior during extension might be modeled as series of elastic springs with a viscous element corresponding to the unfolding of the individual titin domain. The number of force peaks arising in the profiles is equal to the number of 1tit domains involved in the stretched protein. As shown in Fig. 3.18, titin generates a restoring force based on the mechanism of entropic elasticity. The elasticity of entropic components allows titin to be extended fully reversibly at physiological forces, without the need to unfold the "1TIT" domains. The elastic component arises from the extension of the serially linked "1TIT" domain chain. The continuum nanomechanical model can be described as following.

Force Stretching Modeling

In continuum mechanics, the stretching force with respect to the elongation $F(z)$ is given by the following relationship:

$$F(z) = \begin{cases} f_{WLC}(z) - f_1(z) & \text{if } z < L, \\ f_{ENG}(z) + f_2(z) & \text{if } z \geq L, \end{cases} \tag{3.7}$$

where $f_{WLC}(z)$ and $f_{ENG}(z)$ denote the entropic elasticity and tensile force. The boundary conditions are as follows:

$$f_1(z) = 0 \quad \text{for } z \to 0, \qquad f_2(z) = 0 \quad \text{for } z \gg L, \tag{3.8}$$

$$f_1(z) \to \infty \quad \text{for } z \to 0, \tag{3.9}$$

$$f_2 \text{ reaches a finite value for } z \rightarrow L. \tag{3.10}$$

At low stretch forces, the titin molecule was proposed to act as an "entropic spring" showing a worm-like chain behavior. Taken simply, it describes a molecular chain as deformable continuum or rod of given persistence length, A, which is a measure of the molecule's stiffness. The relationship between the end-to-end length (z) and the external force (f) are given by the curve (Fig. 3.20):

$$f_{WLC}(z) = \frac{kT}{A}\left[\frac{1}{4(1-\frac{z}{L})^2} - \frac{1}{4} + \frac{z}{L}\right] \tag{3.11}$$

where k is the Boltzman constant, T is the temperature, and L is contour length (the length of fully extended WLC chain). For our simulations the persistence length is settled to $A = 0.4$ nm [50]. In order to determine the value of the protein length L, we summed the contour lengths of all the "1tit" domains when considered in its fully extended state. It results that the contour length is proportional to the number of amino acids [51].

The tensile force between two Ig proteins is due to hydrogen bonding which can be expressed as:

$$f_{ENG}(z) = k_T(z)(z_{break} - z) \tag{3.12}$$

with

$$k_T = H(z_{break} - z)\begin{cases} k_T^{(0)} & \text{if } z < z_1, \\ k_T^{(1)} & \text{if } z \geq z_1, \end{cases} \tag{3.13}$$

where $k_T^{(0)}$ and $k_T^{(1)}$ are spring stiffness constants optimized by quantum mechanics.

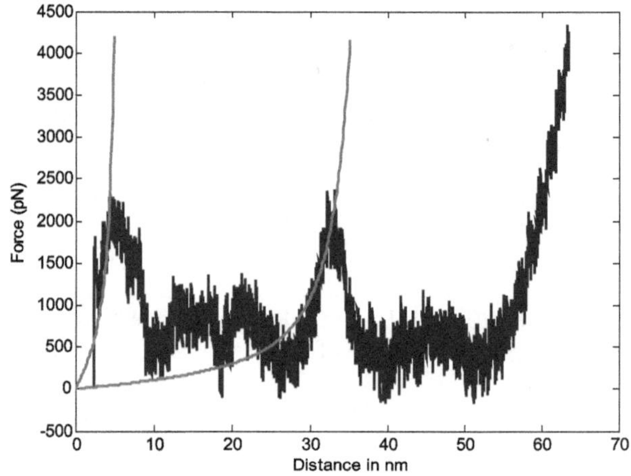

Fig. 3.20 Entropic force-distance relation of elastic region of titin modeled as two serially linked WLCs

Force Relaxation Modeling

The stress relaxation data were fitted to a three-order exponential decay functions of the following type:

$$f(t) = f_0 + A_1 \times e^{\frac{-(t-t_0)}{t_1}} + A_2 \times e^{\frac{-(t-t_0)}{t_2}} + A_3 \times e^{\frac{-(t-t_0)}{t_3}} \quad (3.14)$$

where f_0 and t_0 are offset and center, respectively, A_1 through A_3 are decay amplitudes, and t_1 through t_3 are decay time constants. Curve fitting was done by using a nonlinear least-squares method.

Based on these continuum models, the tandem-Ig proteins are modeled by WLC models in molecular dynamics without changing the graphite monolayer model (Fig. 3.22). The WLC model is implemented as TCL script, and communicate in realtime with NAMD program during simulations. In order to compare the performances of quantum mechanics-molecular dynamics (QM-MD) and quantum mechanics-molecular dynamics-continuum mechanics (QM-MD-CM), we performed several calculations (see Table 3.2). It can be noticed that for acceptable results of simulation (Fig. 3.21), the QM-MD-CM reduces the complexity of nanorobotic structures (decrease of number of atoms) to be simulated for smaller computational time (around few picoseconds) and smaller simulation times (nanosecond order).

Finally, results of stretching-relaxation curve for both types of simulations, i.e., QM-MD simulation (Fig. 3.23) and QM-MD-CM (Fig. 3.24) show a good reversibility of the protein-based nanorobotic platform motion when completely relaxed. The error tends to be lower (less than 6 percent) in the case multiscale QM-MD-CM simulation for small stretching deformations of the protein-based nanosprings.

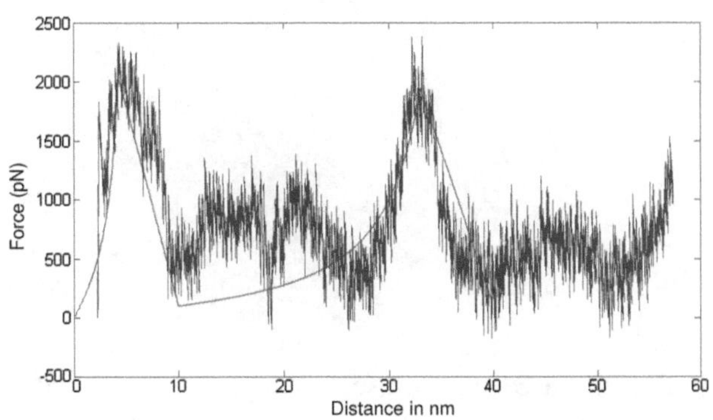

Fig. 3.21 Force-extension curve for molecular dynamics-continuum simulations (QM-MD-CM)

Table 3.2 Performance comparison

Simulation	Nb atoms	Simulation time	Computational time
QM-MD	9984	56×10^{-9} s	52.56×10^{3} s
QM-MD-CM	1718	13×10^{-12} s	58.052×10^{-9} s

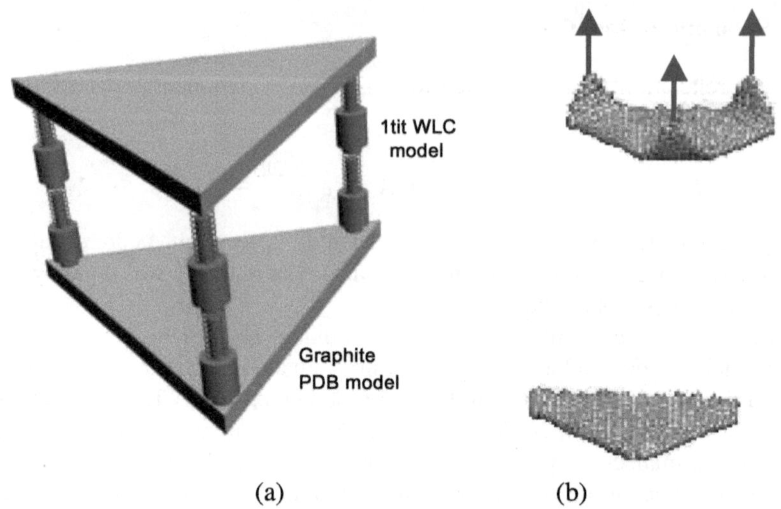

Fig. 3.22 Multiscale nanorobot simulation. (**a**) The graphite was modeled by PDB model and the titin by WLC model, and (**b**) multiscale simulation using continuum, molecular and quantum dynamics (the passive links are not represented for convenience)

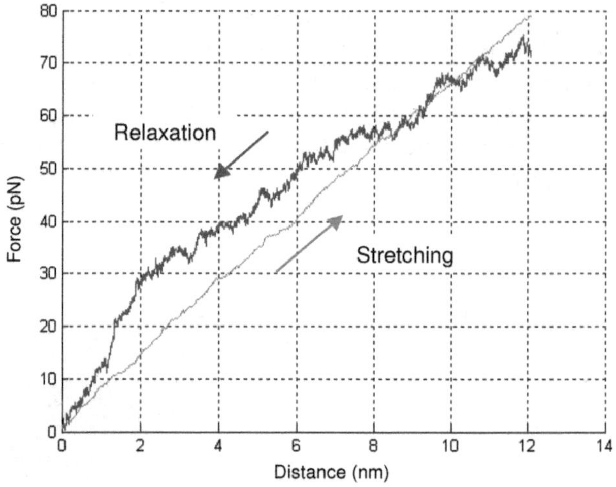

Fig. 3.23 Force-extension curve when the nanorobotic platform is stretched and relaxed. Simulation performed using QM-MD modeling

Design based on multiscale coupled with virtual reality advanced techniques approach is very promising in the domain of bio-nanorobotics. We have demonstrated that realistic design and simulations can be carried out by integrating the physics at several scales (atomistic, nanometer, molecular and mesoscopic) for various times scales of simulation (from few nanoseconds to several milliseconds). The results shows that reduction of 10^6 seconds in computation time with very short time scale length of simulation 10^{-12} seconds can be obtained using multiscale modeling (molecular dynamics and continuum mechanics) approaches.

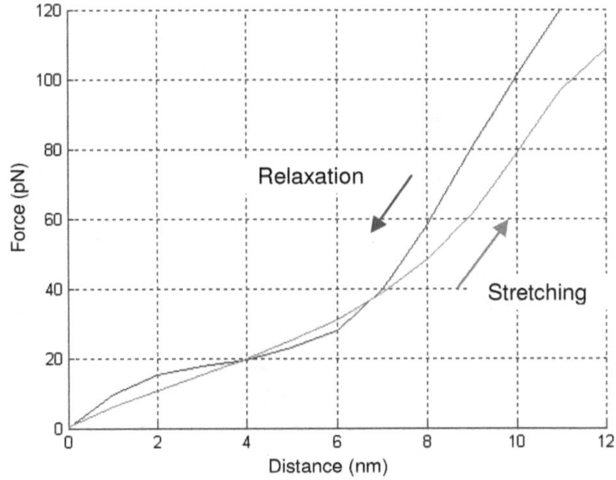

Fig. 3.24 Force-extension curve when the nanorobotic platform is stretched and relaxed. Simulation performed using multi-scale QM-MD-CM modeling

Comparison with Literature Experiments

The experimental investigation of molecular sarcomere elasticity mechanism driven by myosin-actin microfilaments is a challenging research issue. Most of the information about single molecule mechanics of motor proteins arises from the *in vitro* gliding and bead assays during which the proteins are submitted to mechanical forces. These methods allow for the manipulation of single molecules, to observe single events in a chemical reaction like ATP hydrolysis, and to measure mechanic parameters during stretching, and force generated by single molecules [10]. These techniques do not allow one to find out how different types of deformation are correlated with protein structure. Moreover the experimental techniques usually employed for motor proteins characterization have one limitation consisting in the direct observation occurring during mechanical tests. Due to the small scale structure of the proteins it is difficult to anchor them, assemble them, and control their behavior during the pulling experiments. A quantitative analysis of this is beyond the scope of the present work but the magnitude of the forces required to stretch titin repeats can be experimentally and theoretically estimated as a proof-of-concept. Micromechanical studies on isolated single titin molecules recently have demonstrated that the Ig domains and PEVK segment can unfold at high forces and thereby contribute to the extensibility of the sarcomere [41]. To understand how it is possible that titin's distinct nanomechanism extends, we simulated the segmental extension with the multiscale QM-MD-CM methodology. A double tandem 6 Ig27 repeats composed of 154476 atoms in a solute water medium (pH $= 5.5$, $T = 300°$ K) has been thermalized and solvated with the TIP3P model for water before to apply stretching forces. During the first simulation steps, the poly-Ig chains lengthen mainly within a low SL range where passive force is very small.

The first stretch would involve straightening out the tandem-Ig, disrupting the interfaces between adjacent domains, which are relatively small areas of contact,

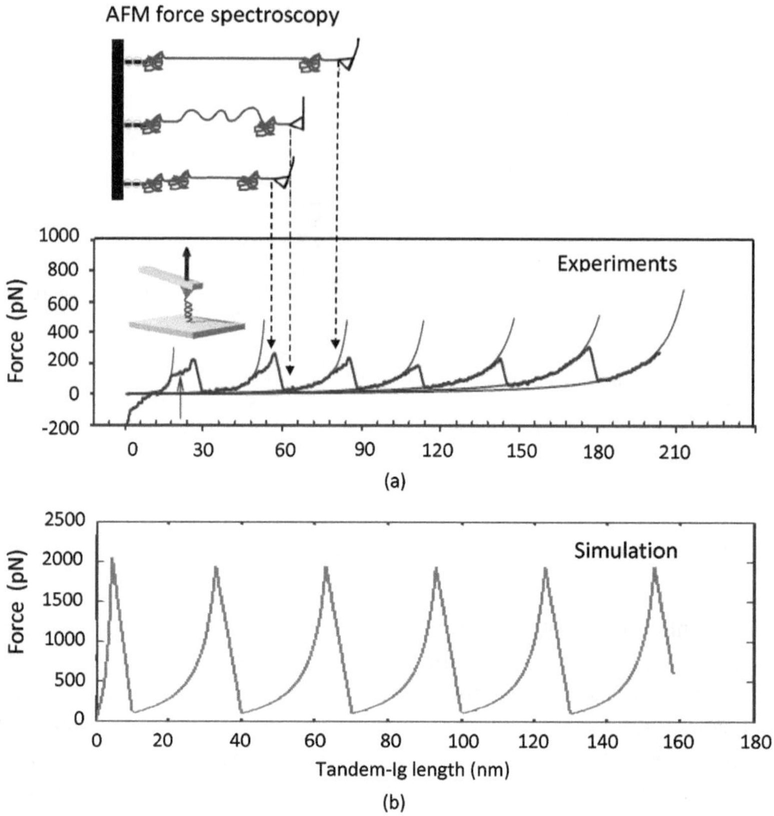

Fig. 3.25 (**a**) AFM force spectroscopy of a heteropolyprotein containing six I27 Ig domains: Representative force-extension curve obtained by stretching a single-molecule containing I27 Ig domains obtained with permission from [49]. (**b**) Simulation results of a heteropolyprotein containing six I27 Ig domains

equivalent to weak protein-protein interaction. The second step of stretching simulation involves unraveling I27 Ig domains.The tensile force on titin operates on the N- and C-termini of all domains. A typical experimental force-extension curve of Ig I27 is shown in Fig. 3.25(a) through AFM force spectroscopy obtained from the data provided by the authors in [49]. A single molecule is identified by the correct spacing between force peaks (the countour-length increase, ΔL, measured by the WLC model, is 28.1 nm for one unfolding event) and the unfolding force magnitude of $\sim$200 pN. The initial part of the force-extension curve is fitted with the WLC model to obtain the entropic spring parameters. The simulated results using the proposed approach is shown in Fig. 3.25(b). The average contour length of a fully extended tandem-I27 Ig repeats fits the real values $\Delta L = 31.6$ nm. However, the mismatch between the real and simulated peak force values are due to the pulling speeds adopted in the simulations that were six to eight orders of magnitude higher than those in the experiments ($\nu = 0.01$ Å/ps). Another typical feature of viscoelasticity is force hysteresis, usually observable as higher force during stretch than during release. We can compared with results of stretching-relaxation curve for both types of simulations, i.e., QM-MD simulation (Fig. 3.23) and QM-MD-CM (Fig. 3.24) for small extensions. The simulation results show a good reversibility of the extensible protein-based nanomechanism when completely relaxed. The error tends to be reduced (less than 6%) for small stretching deformations of the protein-based nanosprings. At high extensions, the QM-MD-CM modeling reveals to be effective in predicting the force hysteresis of the sarcomere mechanism with an errors less than 12 percent. These simulation results are in good agreement with those experimented by the authors in [52] for small deformations. We considered other complex systems with various kinds of forces to testify the performance and complement the approach. Similar results have been obtained on different reversible protein-based nanomechanisms such as four α-helix ROP protein as a parallel spring nanomechanism and a tandem of four double-strand DNA proteins. Furthermore, superelastic series structures with high tensile strength have been experienced, i.e. a pair of double α-helix tropomyosin filaments and β-sheet silk proteins. It allowed to predict the type of force spectra, reversibility, hysteresis, viscoelasticity and irreversible work that may be expected from molecular nanomechanisms. However, the author although validate the one-dimensional molecular link, did not validate the parallel platform for various directional SMD forces. It implies the introduction of additional SMD simulations, i.e., shearing, bending and compression tests before to propose a complete molecular nanomechanical model.

4 DNA Nanorobotics

Problem Formulation

Proteins represent fertile territory for nanoscale machines that produce linear motions in liquid environments. These molecular motors are proteins such as kinesin [20], myosin [18] and dynein [19] that move along tracks (actin filaments, microtubules, etc.) inside a cell. The functioning of these molecular motors is dependent on adenosine triphosphate (ATP), a molecule with high chemical energy. These ATP based protein motors can generate mechanical forces in the range of 10–20 pN depending on the applied load. Several other non-ATP based motors have also been proposed such as viral protein linear motor [53] or GCN4 leucine zipper based nanoActuator [8]. However, due to relatively complex conformational flexibility and free energy instability in the design of protein based tailored bio-nanosystems, we prefer to investigate in this work DNA based nanodevices. DNA undergoes substantial conformational changes in responses to environmental stimuli (temperature, acidic concentration, salt, ionic level) which facilitate controlled mechanical motion compared to proteins. Unusual DNA motifs [23, 54] can be used to construct molecular building blocks by virtue of the fact that the sticky-ended association of DNA molecules occurs with very high specificity. Branched DNA molecules with sticky ends are promising for assembling robotic based-DNA nanostructures. Simmel [55] reported the construction of a mechanical DNA-based device that might serve as the basis for a nanoscale robotic actuator. The mechanism has two rigid double-stranded DNA arms a few nanometers long that can be made to rotate between fixed positions by introducing a positively charged cobalt compound into the solution surrounding the molecules. Yan et al. [56] proposed a X-shaped DNA tiles linking a square grid with some DNA strands that can lengthen or shorten few nanometers like tiny pistons. In the following sections, we proposed different computational studies to demonstrate a general approach towards the design, computational analysis and characterization of a controlled DNA based nanoactuators, e.g., linear nanotube servomotor and controllable nanogripper.

5 Design and Computational Analysis of a Linear Nanotube Servomotor Using DNA Actuation

We propose in this section a novel concept of encapsulated DNA molecule acting as nanoscale actuator inside carbon nanotubes in a water solute environment. We report molecular dynamics simulations of the dynamic processes towards the prototyping of biological servo nanoactuators (termed, DNA@MWNT). The results indicated spontaneous insertion and confinement of double-stranded Z-DNA molecule under a combined action of van der Waals and hydrophobic interaction forces. Under the temperature-dependent conformational relaxation of DNA encapsulated in a double-walled carbon nanotube, a controllable and reversible linear motion has been investigated using molecular dynamics simulation. Dynamics atomistic-continuum modeling of the forces and energies involved in the driving mechanism has been investigated in order to optimize its displacement-force characteristics. To improve the precision of DNA@MWNT nanoactuator, *in-situ* position biosensing feedback in water environment is necessary. Molecular dynamics simulations revealed the molecular transport dynamics of single-walled carbon nanotubes (SWNT) channels conducting water [57], ions [58] or nucleic acids [59]. Based on these biological channels concepts, a current-based position sensing system is made thus possible. We investigated a new ionic position feedback through the dependence of chlorure ion diffusion (Na^+) by thermal fluctuations on interelectrode distance. The results paves the way for future applications of linear nanotube servomotor acting as a controlled miniature needle for selective cancer cell destruction.

Design of Encapsulated DNA-Double Walled Carbon Nanotube Actuator (DNA@MWNT)

Design Concept

The discovery of multi walled carbon nanotubes offers many scientific challenges for their adaption to a wide variety of nanometer scale actuators. The low-frictional

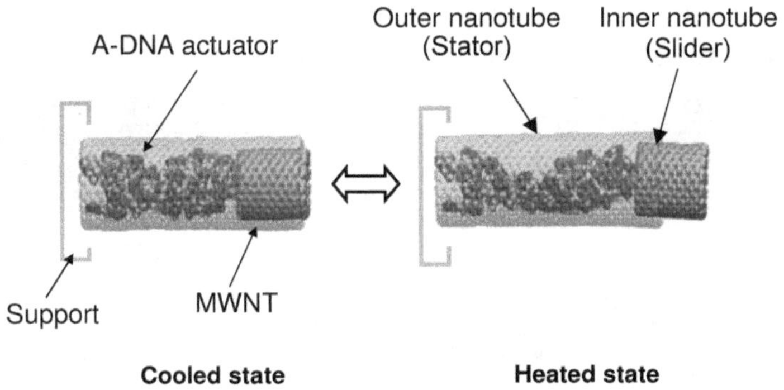

Fig. 3.26 Basic concept of a linear encapsulated DNA-MWNT nanoactuator controlled by temperature parameter

effect for a sliding inner tube inside multi-walled carbon nanotubes have been exploited to create gigahertz oscillation [60], rotational actuators with a nanotube as rotation bearing [61] or telescoping MWNT linear nanomotors using field emission [62]. Figure 3.26 shows a new concept of open ended double-walled carbon nanotube actuator working in a biological medium. In the proposed design, the sliding inner tube is actuated by an encapsulated dsDNA connected to a support electrode. The structure is like an hairpin composed of two coils, having each C-terminal connected to the sliding inner nanotube. The inner nanotube slides when undergoing a conformational change induced by temperature variation (increase or decrease) in mildly acidic medium (pH = 5.5). The design optimization and performance analysis of the DNA@MWNT actuator performances is performed by design software with molecular computational package.

Choice of DNA Actuation

Chemically, DNA is a long polymer made up of a linear series of subunits known as nucleotides. Structurally, DNA molecule is usually found as a double helix, with two strands wrapped around one another. However, DNA molecule can adopt other configurations and can exist in single-stranded forms. As shown in Fig. 3.27, three crystallized states of DNA have been studied: (a) A-DNA, (b) Z-DNA and (c) C-DNA.

We studied the mechanical properties of the three models in order to choose the best candidate for the proposed nanoactuator structure. We use a modified steered molecular dynamics (SMD) technique to measure the motive force produced by denaturation of the DNA. to perform this measure we run a molecular dynamics simulation of the DNA denaturation with an applied external constraint to the DNA terminus. This constraint was applied in the form of a harmonic spring of known stiffness k, attached to the center of mass of the terminus nucleic acid. The harmonic guiding potential and the corresponding exerted force for this system are of the form:

$$U = -k(x - x_0)^2/2, \qquad F = k(x - x_0).$$

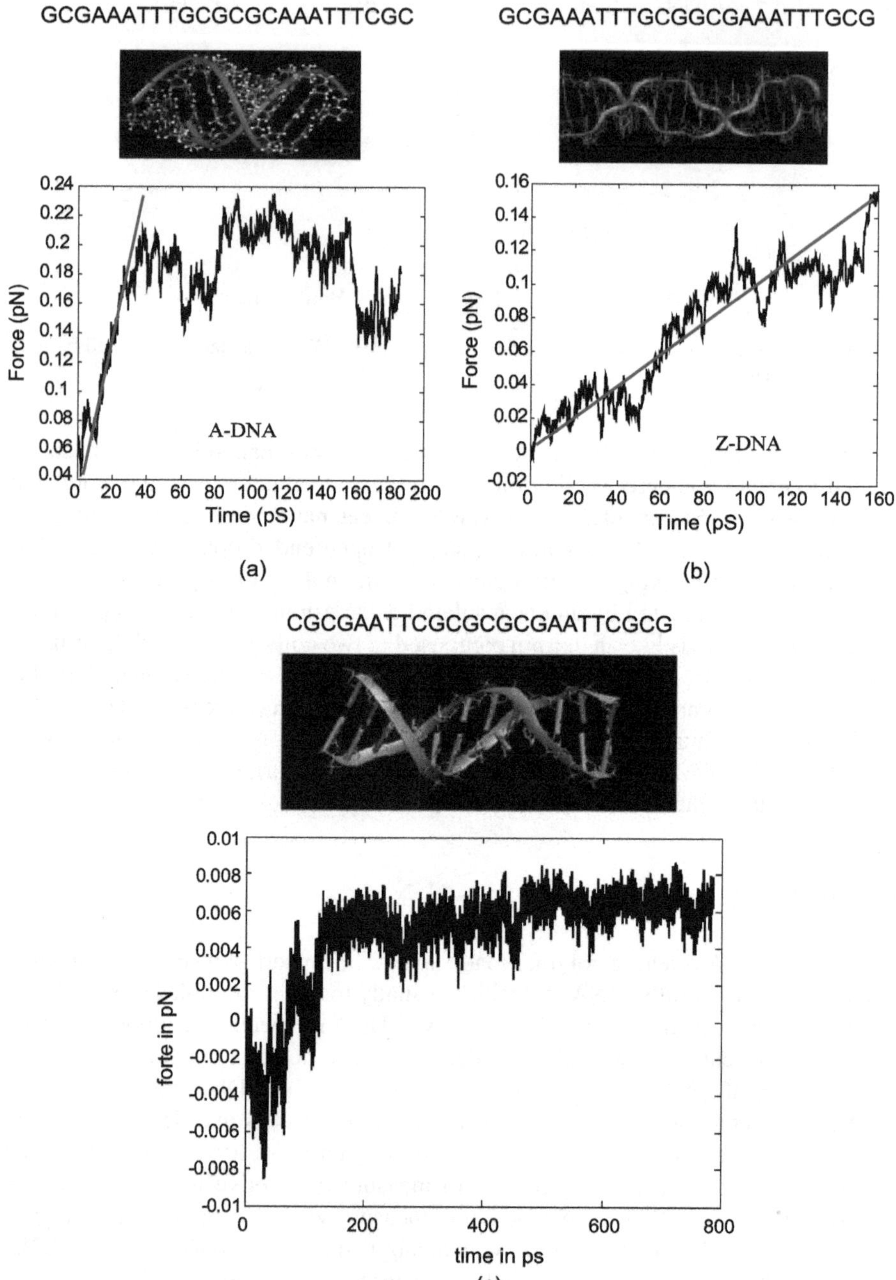

Fig. 3.27 Chemical structure and mechanical force delivered by DNA denaturation for (**a**) A-DNA, (**b**) Z-DNA structures and (**c**) B-DNA

Table 3.3 Characteristics of DNA actuation when denatured by temperature

DNA model	F_{max} (pN)	D_{max} (nm)	T (K)
A-DNA	0.234	0.81	345
B-DNA	0.0082	0.02	337.2
Z-DNA	0.16	1.81	338

Table 3.3 shows the main DNA characteristics with respect to the maximum force F_{max}, the maximum displacement D_{max} and the melting temperature point T. First, we notice that the melting temperature point leading to the opening of dsDNA branchs is very similar. Secondly, the A-DNA molecule presents larger force slope during denaturation compared to Z-DNA and B-DNA molecules but it is limited by its small denaturation displacement. Third, the motive force produced by denaturation of the B-DNA is too small in order to counteract the sum of the interlayer van der Waals interaction (f_{vdw}), electrostatic force (f_{elec}) and the total intershell sliding resistance force (f_r). Finally, the Z-DNA molecule is chosen as nanoactuator with respect to its powerful and controllable driving performances.

Encapsulation of DNA Inside Carbon Nanotubes

Here we report molecular dynamics simulations of the dynamic processes of encapsulating DNA inside CNTs in a solute environment. The nanotube/DNA interaction experiences strong attractive force from each other when interesting their separation is about 1 nm [57]. On this basis, we investigated a DNA-encapsulated SWNTs procedure when applying simultaneously direct current (DC) and radio frequency (RF) electric fields to a substrate coated with open-ended SWNTs. Such non-organic interaction occurs when immersed in an electrolyte plasma. By Raman spectroscopy and HR-TEM analysis, the encapsulation of DNA inside SWNTs is enhanced when both the DC and RF electric fields are simultaneously superimposed. These results indicate that a process of superimposing an RF electric field upon a DC electric field plays a decisive role in the DNA-encapsulated SWNT formation in this solution phase procedure such as in the electrolyte plasma [63]. We simulate the molecular dynamics of the DNA encapsulation inside a SWCNT. A direct electric field is applied to the electrolyte plasma containing DNA negative ions in order to irradiate the single-walled carbon nanotubes with DNA ions. The designed DNA-CNT system consists of a homogeneous single-strand DNA oligonucleotide with 8 adenine bases and an uncapped armchair (16, 16) carbon nanotube (5.84 nm long and 2.14 nm in diameter). As initial configurations, CNT and DNA were aligned along the nanotube axis and separated by 0.65 nm. The CNT-DNA complex was solvated in a water reservoir and its dynamics was simulated for 2 ns at temperature 355 K. The nanotube charge distribution on the single-walled carbon nanotube is obtained by an atomistic moment method based on classic electrostatics theory. A time-step of 2 fs was used and full-precision trajectory was recorded every 1 ps. The snapshots of the oligonucleotide-nanotube system shown in Fig. 3.28 indicated a very

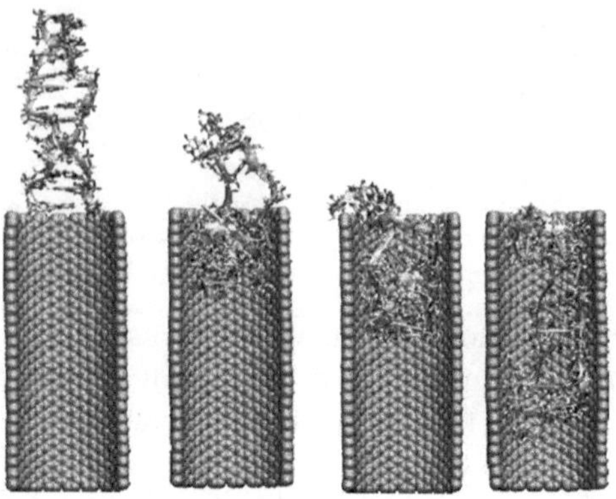

Fig. 3.28 Simulation snapshots of a DNA insertion inside SWCNT. Water molecules are not displayed for clarity

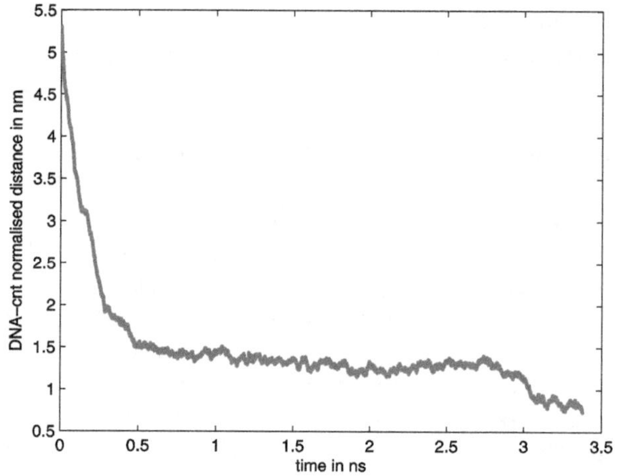

Fig. 3.29 Normalized center-of-mass distances between the oligonucleotide and carbon nanotube as function of simulation time

fast insertion process of oligonucleotide into nanotube. These simulation results are similar to those of Gao et al. [64]. At $t = 30$ ps, the first base of the oligonucleotide has begun to enter the nanotube. After 500 ps, five of the eight DNA bases are fully inside the nanotube and the first base has reached the opposite end of the tube. The derived van der Waals energy between the nanotube and the first DNA base entering the nanotube decreases greatly with distance. Correspondingly, the center of mass distance between oligonucleotide and carbon nanotube rapidly decreases with time up to 500 ps (see Fig. 3.29). As shown in Fig. 3.30, the non-bonded energy interaction increases greatly with respect to the DNA insertion due to the strong attractive van der Waals interaction. At the equilibrium point, the DNA is in a stable sustentation state inside the nanotube.

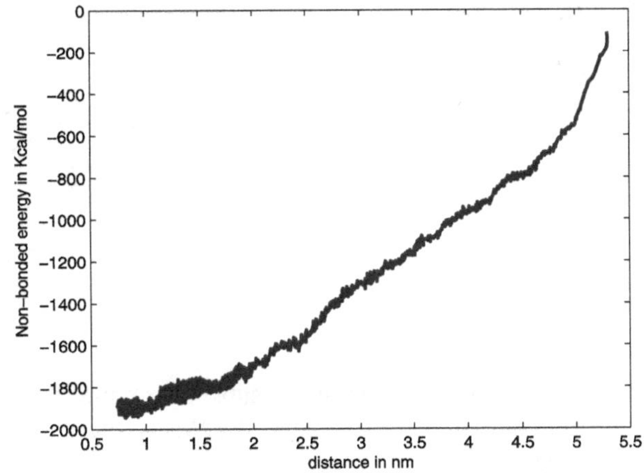

Fig. 3.30 Non-bonded energy interaction between the oligonucleotide and carbon nanotube as function of center-of-mass distance

DNA@MWNT Nanoactuator Modeling

Multiscale Approach

The limitations of purely atomistic or purely continuum simulations, have motivated research in multi-scale simulations that bridge atomistic simulation and continuum modeling. In order to make the computations selective, multi-scale models usually decomposes the domain into coarse/fine subregions. We have developed a design and modeling platform that couple different frameworks. We use several physics for particular length and time scales to characterize and treat aspects of nanostructure phenomena that operates only over those scales. Through quantum mechanics calculation, we optimized the geometry of the single walled CNT. As the dsDNA is attached to single walled CNT by -COOH- group a novel nonstandard group becomes created and must be parameterized. By using Density functional theory method we calculated the bond length, angles, electrostatic potential (ESP), charges and vibrational frequencies of theses linkage groups. These parameters are then used at the molecular dynamics level for nanoactuator optimization.

Atomistic and Molecular Modeling of DNA Encapsulated CNTs

The dynamics of the slider carbon nanotube (CNT) can be represented by the scheme of Fig. 3.31. By assuming the dynamics along the z-axis, the Newtonian mechanics equation can be written in a steady-state:

$$f_{DNA} + f_{elec} - f_{vdw} - f_{sub} = 0 \tag{3.15}$$

where f_{DNA} represents the force delivered by DNA during denaturation, f_{elec} represents electrostatic repulsive force between the stator CNT and slider CNT, f_{vdw} represents the van der Waals forces between the two CNTs and f_{sub} represents the

Fig. 3.31 Dynamics model of the nanomotor actuated by DNA molecule

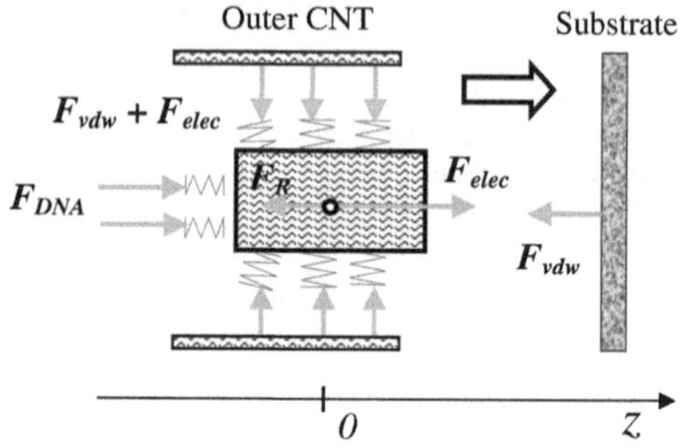

non-bonded forces applied by the substrate on the slider carbon nanotube, this force is the total of electrostatic and van der Waals interactions. We assume to be negligible the mechanical interlayer sliding force f_r.

The Electrostatic Force

According to the Coulomb's law, the resultant electrostatic force between the two cnt's can be calculated by the following equations:

$$f_{elec} = \sum_{j=i+1}^{n} \frac{q_i q_j}{4\pi\epsilon_0\epsilon_r\delta_{ij}^2} \tag{3.16}$$

where and q_i and q_j are magnitudes of the charges of two nucleotides; δ_{ij} is their separation distance; ϵ_0 is the permittivity of free space; and ϵ_r is the relative dielectric constant of the medium in which the charges are placed. The line of action of the electrostatic force is assumed to be along the direction of motion of the inner tube applies.

The Van der Waals Force

The van der Waals force is a short-range force caused by instantaneous dipole interactions. The force is comparatively larger in air than in a liquid media, and is proportional to the Hamaker constant, which is one of the leading parameters to accurately estimate the van der Waals forces. The van der Waals's forces is included by using the Lennard-Jones potential as:

$$f_{vdw} = \sum_{excl(i,j)=1} \epsilon_{ij} \left(\left(\frac{R_{ij}}{r_{ij}} \right)^{12} - \left(\frac{R_{ij}}{r_{ij}} \right)^{6} \right) \tag{3.17}$$

where ϵ_{ij} represents the energy of the minimum (deepest) point on van der Waals curve for the atom pair i–j, R_{ij} is the separation distance between the atom pair i–j at the energy minima, and r_{ij} is the current actual distance between atom i and atom j. The attractive forces between two proximal atoms are proportional to r_{ij}^{-6} and the repulsive forces due to their nuclear repulsions are proportional to r_{ij}^{-12}. Figure 3.32(a)–(b) visualizes the simulation characteristics of the forces involved in the driving mechanism. The set of curves in Fig. 3.32(a) shows the linear region of control of the nanoactuator. The DNA force F_m increases gradually as the sliding nanotube is moving until to saturate. It should be noticed that the electrostatic repulsive force F_{elec} decreases with respect to the slider motion. On contrary, the van der Waal force f_{vdw} present two distinct states: repulsive interaction and attractive interaction. The set of curves in Fig. 3.32(b) shows the force variation when the DNA actuation is saturated. Close to the substrate, we can see a drastic decrease of the electrostatic repulsive force and van de Waals interaction. It should be noticed that the nanoactuator is fully reversible as shown in Fig. 3.32.

Servo Nanoactuators Using Ionic Current Feedback

To improve the precision of DNA@MWNT nanoactuator, *in-situ* position biosensing feedback in water environment is necessary. Molecular dynamics simulations revealed the molecular transport dynamics of single-walled carbon nanotubes channels conducting water [57], ions [58] or nucleic acids [59]. The activation energy barrier for entry of ions through hydrophobic carbon nanotubes is caused by the fact that water molecules being immobilized inside the tube would require considerable energy to reorient them around the ions as they do in the bulk. This energy corresponds to the free energy of solvation allowing ion permeation [58]. This additional energy can be provided by (1) an external electric field or (2) the presence of charged atoms on the nanotube. As shown in Fig. 3.34, we investigated the former solution in order to simulate biological current sensor. We investigated a new ionic position feedback through the dependence of sodium ion diffusion (Na^+) by thermal fluctuations on interelectrode distance. Even though an electrical field alone would drive ions into the nanotube, the partial charge on the substrate increases strongly the sensitivity of permeation. Furthermore, it can be used to control the rate of the ionic flow into the slider nanotube. An electric field of $0.15\ \text{V/nm}$ was used to drive the ions through the CNTs.

Figure 3.35 shows clearly some snapshots of permeation of (Na^+) ions and water molecules through the double walled inner CNT. It can be explained by the fact that when the temperature increased during denaturation of Z-DNA, the thermal fluctuation of the nanochannel increased, the rate of ion injection into the nanochannel slightly increased and then, the injected ions easily moved toward the other side without disturbance. A closer view shown in Fig. 3.36 visualizes the directional water and ions flow under a hydrostatic pressure difference induced through the application of attractive electrostatic force. The occupancy of ions in the tube depends strongly of the current position of the slider nanotube.

To characterize ion transport through the nanotube for a given electric voltage V, we calculated the electrolytic current I_e (from the ionic electric charges q_i) as a

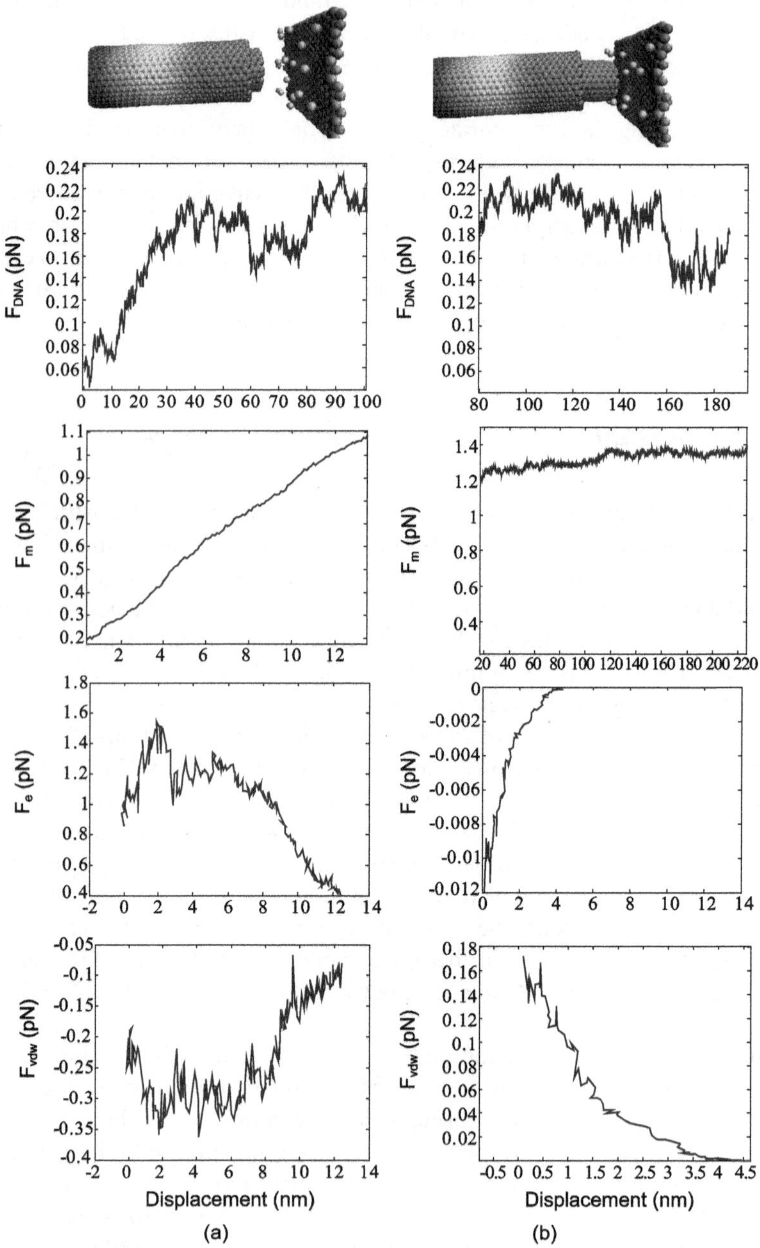

Fig. 3.32 Force balance of the nanoactuator. (**a**) Forces when the inner CNT is far from the substrate and (**b**) forces when the effector is closed to substrate

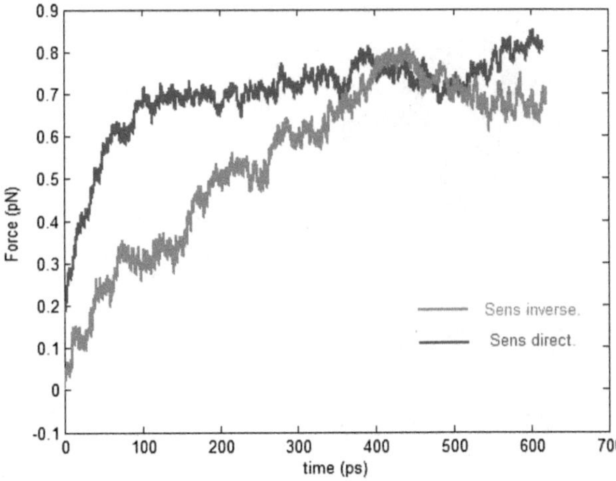

Fig. 3.33 Motion force for a forward and backard motion

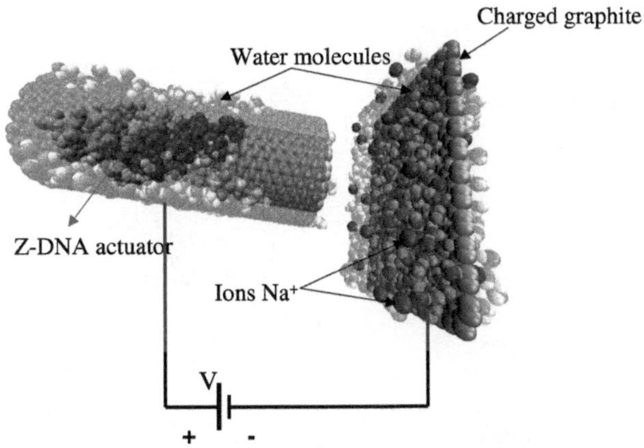

Fig. 3.34 Structure of the servo nanoactuator with ionic current feedback for position measurement

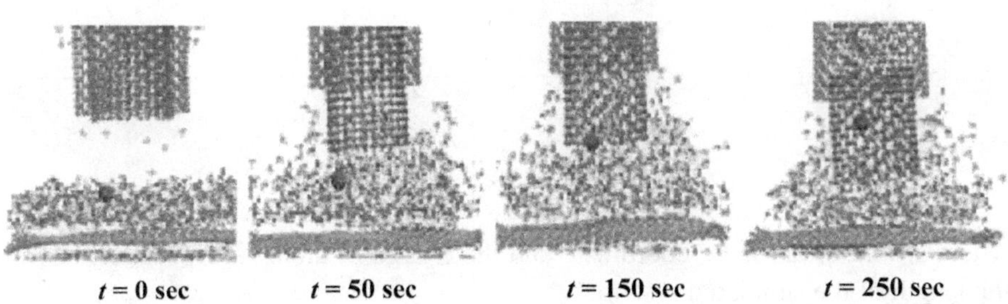

Fig. 3.35 Snapshots of permeation of Na^+ ions through the double walled carbon nanotube. Water molecules are also shown during permeation

function of time t. The current expression is given by:

$$I_t = \frac{1}{\delta_t L_z} \sum_{i=1}^{n} q_i (z_i(t + \delta_t) - z_i(t)) \tag{3.18}$$

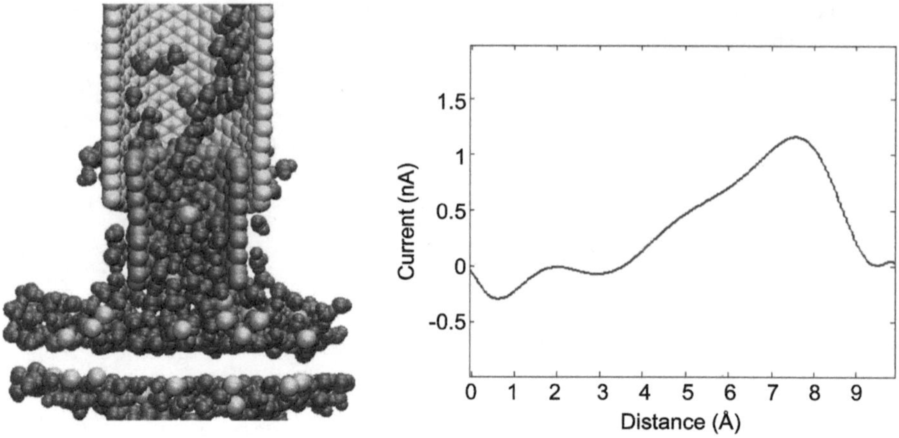

Fig. 3.36 Position trajectory of Na^+ ions during transport

Fig. 3.37 Current-time characteristic

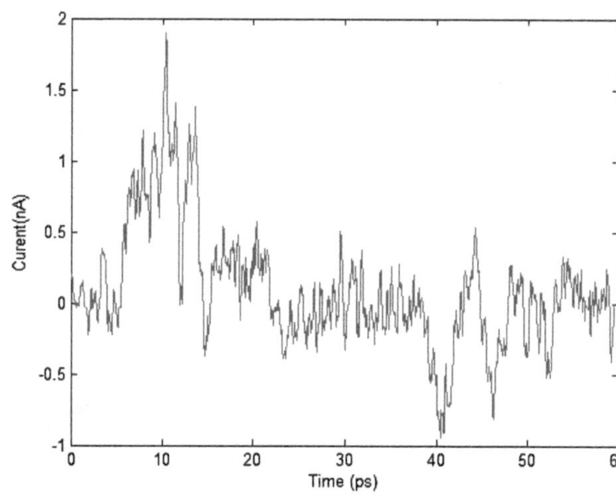

where z_i and q_i are the z coordinate and the charge of atom i, respectively; L_z is the length of the simulated system.

By measuring the ionic current, we can achieve an efficient position control of the CNT effector as shown in Fig. 3.37. It should be noticed that the ionic position sensor is quasi-linear in the range of 3 Å to 8 Å. Furthermore, we can notice that current decreases (after $t = 10$ ps) as the sliding tube approaches the substrate. By filtering the ionic current, a linear current-distance calibration has been simulated. Further work on the implementation of the ionic current sensor in a position feedback system is under investigation.

Force Control of Nanoscale Servomotor Using Variable DNA Stiffness

The modulation of the strength of the DNA based actuator has been considered. In this regard, force control using variable DNA stiffness can be envisaged since previous experiments on DNA overstretching have shown that effects of solution conditions plays an important role in force-extension curves. Variations of the medium environment have been controlled through three parameters: (i) the temperature, (ii) the level of acidic pH and (iii) the electron density of the molecule.

The strong sensitivity of the force-extension characteristics of the DNA molecule to temperature plays an important role in the control of the DNA stiffness [65]. Indeed, the overstretching portions of the force-extension curves as a function of temperature decrease greatly as the temperature is increased. As example, Fig. 3.38(a) shows the evolution of the DNA spring constant K_{DNA} with respect to the temperature ranging from 273 K to 450 K. In theory, the results show clearly the linear decrease of spring constant with respect to the temperature. In practice, for such high temperatures the DNA becomes completely denaturated (DNA melting) and reversibility is alterated by the strong hysteresis. Removal of the bonds separating the melted base pairs represents an irreversible process, in which the two strands completely separate. The blue dotted line of Fig. 3.38(a) shows this irreversible process. In a realistic way (red solid line), the adjustable stiffness can be adapted between 780 N/m to 720 N/m in relation to the temperature control (ranging from 300 K to 338 K). Such results have been confirmed by an experimental investigation carried out M.C. Williams et al. in [66]. The authors demonstrated that the temperature dependence of DNA stretching in 500 mM ionic strength Hepes buffer at pH = 7.5 exhibited large variations of the DNA stiffness between 284 K to 325 K in a reversible way (see Fig. 5 in [22]). The experiments confirmed the very little hysteresis of DNA molecule with a thermal bath at 300 K. However, it should be noticed that when stretched at 318 K, almost the entire length of DNA that was stretched remains single-stranded as it is relaxed.

Practically, the DNA molecule is able to perform repeatable motion controlled by variation of pH by adding protons (termed protonation). It is proposed to isolate this domain from the DNA and trigger the stiffness change by variation of pH (lowering the melting point of dsDNA). Figure 3.38(b) shows the change of the protein stiffness with respect to the pH value. For a neutral acidic pH value (pH = 11), the DNA protein is in a partially α-helical stranded coiled coil. The value of pH = 5.5 corresponds to the neutral state. An increase occurs at lower pH between 5.5 and 10 leading to a stiffness variation ranging from 670 N/m to 780 N/m. For this configuration, it is necessary to protonate the amino acid side chain of the protein by adding proton inside the native environment. These values are in good agreement with those that have been reported previously by M.C. Williams et al. in [67]. The authors showed that as pH is raised between pH = 6.0 and pH = 10.6, the shape of the force-extension curves changes very little but the value of the overstretching forces decreased greatly explaining the linear decrease of DNA stiffness in Fig. 3.38(b). At low pH levels ranging from pH = 2 to pH = 5.5, the simulations

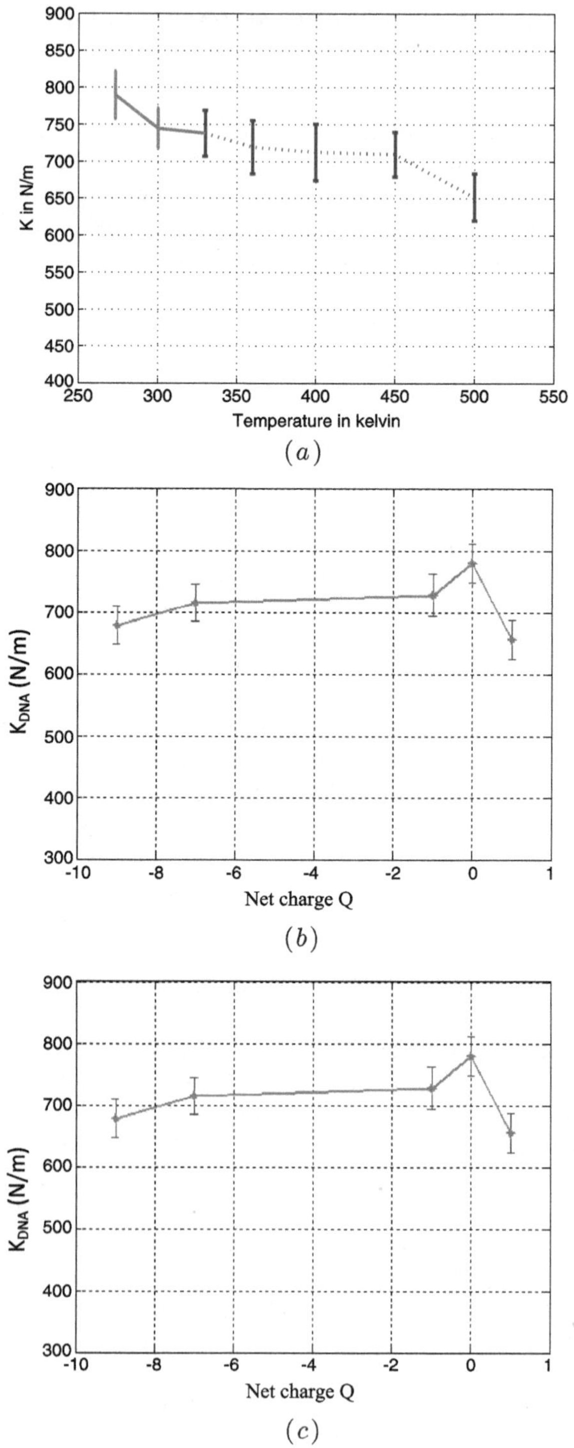

Fig. 3.38 (Color online) Evolution of the spring stiffness of dsDNA during stretching for changing medium properties: (**a**) temperature, (**b**) level of acidic pH and (**c**) level of electron charge Q

showed that the force-extension curves could be reproduced without any hysteresis indicating that the changes to DNA molecule at low pH are fully reversible. these results are in good agreement with experiments carried out in [68].

Finally, the increase or decrease of the electron density (namely, the electric charge Q in Coulombs) has been investigated. The concept of using electrons as a medium for such purposes is not unreasonable because electron density will readjust along the molecular backbone in response to an inflow/outflow of electronic charge. Moreover, using electrons as a mechanism for changing the shape and rigidity of molecular systems has precedence [69]. Figure 3.38(c) shows the change of spring stiffness according to the electron charge of the protein. From these results we derive that the force stiffness bracket a range from 680 $\mathrm{N/m}$ to 780 $\mathrm{N/m}$ for the ds-DNA protein. On contrary to the temperature and pH dependence (Fig. 3.38(a)–(b)) where the base pairs connecting the DNA strands are broken during the overstretching simulations, the ionic dependence (Fig. 3.38(c)) of the overstretching indicates that the two DNA strands remain close together during stretching. This implies that the majority of DNA base pairs melt within internal domains rather than from the free ends. Experimentally one can control the sodium ion concentration $[Na^+]$ of the medium in order to adjust the stiffness of the elastic DNA molecule. Our simulation results are in agreement with the partial stretching curves obtained in salt concentrations ranging from 1000 to 2.57 mM carried out by the authors in [54]. For low level of electron charges (namely, low level of salt concentration), the DNA stiffness increases linearly with respect to the salt concentration $[Na^+]$. As conclusion, double-stranded DNA (dsDNA) molecules have interesting characteristics due to their elastic behavior, reversibility, stability and structural conformation. As example, DNA can provide three independent reversible degrees of freedom: stretching, shearing and twisting. They can be used to construct molecular building blocks by virtue of their conformation pathways.

Controllable DNA-based NanoGripper

Design of DNA Actuation

In this section, we study the molecular properties of DNA molecules to change their 3-D conformation depending on the temperature level of the native medium (water). Thus, a new biomolecular gripper type called dsDNA-SWNT nanogripper is designed (Fig. 3.40). It is composed of a dsDNA protein with two SWNT as nano end-effectors operating in mildly acidic medium (pH $= 5.5$). The structure is like an hairpin composed of two coils, having each C-terminal connected to one SWNT. At unstressed state, the dsDNA is stabilized against spontaneous dissociation into two strands by the interaction of free energy $g_0 = 1.4K_BT$ per base pair. The nanogripper is at rest (closed state). When the DNA molecule is subjected to an external energy (thermal denaturation) greater than the free-energy potential barrier g_0, initiation of unzipping of the two single-strand extremities occurs. The nanogripper

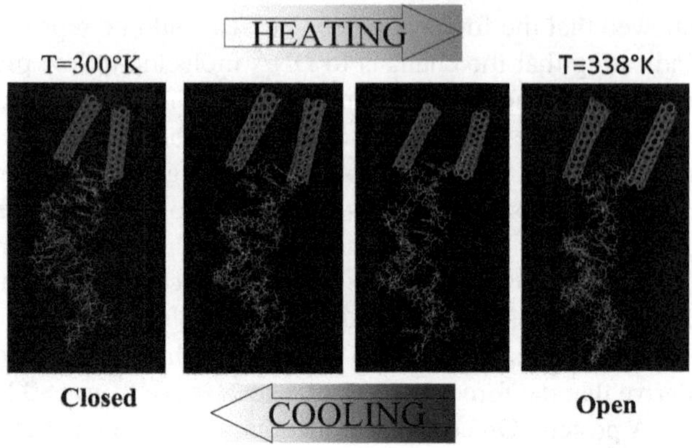

Fig. 3.39 dsDNA-SWNT nanoGripper. In the native state the nanoGripper is closed ($T = 300$ K) and in the conformational state ($T = 338$ K) the nanogripper is open

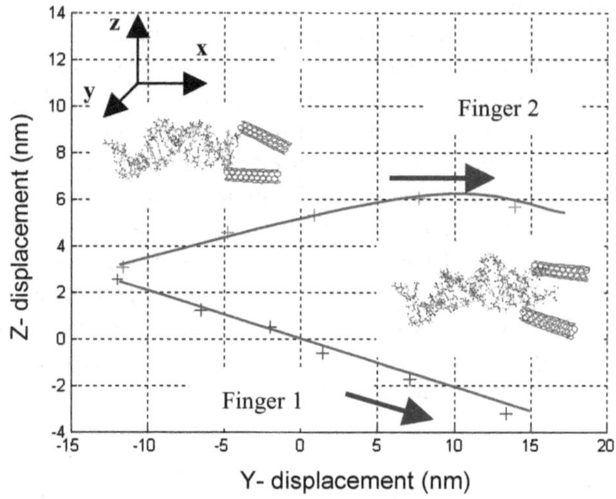

Fig. 3.40 Motion of the nanofingers during elevation of the temperature from native state (nanogripper closed) to conformational state (nanogripper open). Influence of the dsDNA configuration with single walled carbon nanotubes

initiates the opening of the nano end-effectors until to reach its final open-state. The choice of the dsDNA molecule is thus crucial in the design of the nanoscale robotic gripper. It is important to emphasize the physical factors involved in the conformation, namely contributions from base pairing, base pair stacking, hydration and counterion binding. As these non-covalent interactions depend on the specific chemical composition of the strands used, we investigated the study of three DNA hairpins containing specific base-pairs:

1. Double-stranded A-DNA molecule:

$$GCGAAATTTGCGCGCAAATTTCGC$$
$$CGCTTTAAACGCGCGTTTAAAGCG.$$

2. Double stranded Z-DNA molecule:

$$GCGAAATTTGCGGCGAAATTTGCG$$
$$CGCTTTAAACGCCGCTTTAAACGC.$$

3. Double-stranded B-DNA molecule:

$$CGCGAATTCGCGCGCGAATTCGCG$$
$$GCGCTTAAGCGCGCGCTTAAGCGC$$

where A, C, G and T stands for adenine, cytosine, guanine and thymine.

The simulation results shown that all three hairpins have a similar conformation with different melting temperatures: $T_{A\text{-}DNA} = 345$ K, $T_{Z\text{-}DNA} = 337.2$ K and $T_{B\text{-}DNA}$=338 K. However, the calculation of the non-bonded energies demonstrated that the B-DNA molecule necessitates lower energy values during thermal unzipping ($E_{A\text{-}DNA} = -173.797$ kcal/mol, $E_{Z\text{-}DNA} = -249.638$ kcal/mol, $E_{B\text{-}DNA} = -74.3423$ kcal/mol) making it a good candidate for nanogripper actuation. Figure 3.40 presents the adopted double-stranded B-DNA-SWNT nanogripper design. In native state, the nanogripper's fingers are at rest (closed state) for normal thermal bath conditions ($T = 300$ K). When increasing the temperature of the thermal bath ($T = 338$ K), the nanogripper opens its fingers until to reach the conformation state (open state). The cooperative denaturation of the $3'$–$5'$ terminal pair extremities (at positions ~21–24 with GC base pairs) of B-DNA molecule is fully reversible in the time domain of experience as shown in [72]. A reversible finger's displacement motion is achieved when cooling the thermal bath. Both the forward (closed-to-open) and the reverse (open-to-close) transformations of the nanogripper are carried out for two end-point structures of 10 Å. The selected B-DNA protein is able to perform repeatable motion controlled by variation of temperature. The finger's motion are composed of a linear finger opening (y–z plane) connected to carbon nanotubes as nano end-effectors. In order to assess the interaction forces when handling a nanoobject, we simulated a contact force by using two nonlinear spring K_1 and K_2 connected to each end-effector. Figure 3.41 shows the evolution of the gripping force during the opening of the nanotweezers. The contact force is estimated by $F_i = K_i.x_{i,j}$ with $(i, j) = (1, 2)$ for each nanofinger. It should be noticed that when the nanofingers are in closed contact with the nanoobject, the forces provided by the dsDNA are quite linear with the opening displacement x. Saturation limits of gripping tasks are reached for force levels on the order of $F_{\max} = 0.3$ pN. For large opening displacements, the forces are dissimilar due the small parasitic rotation of the fingers. Figure 3.42 shows the motion reversibility of the nanofingers during a temperature cycle time (heating and cooling) from 300 K to 338 K. Although the overall unfolding pathway is independent of temperature cycling, differences are observed from trajectory-to-trajectory during the cooling procedure. The reversibility error is less than 8 percent in the worst case.

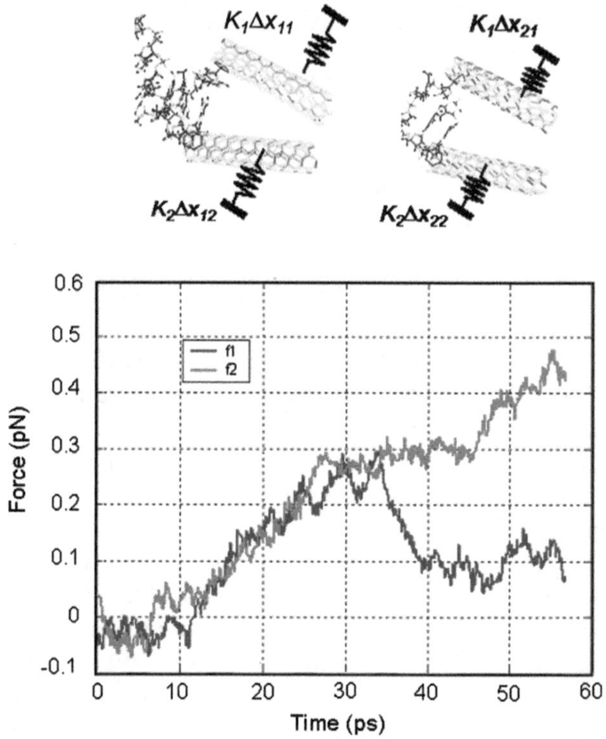

Fig. 3.41 Simulation of gripping forces when handling a nanoobject

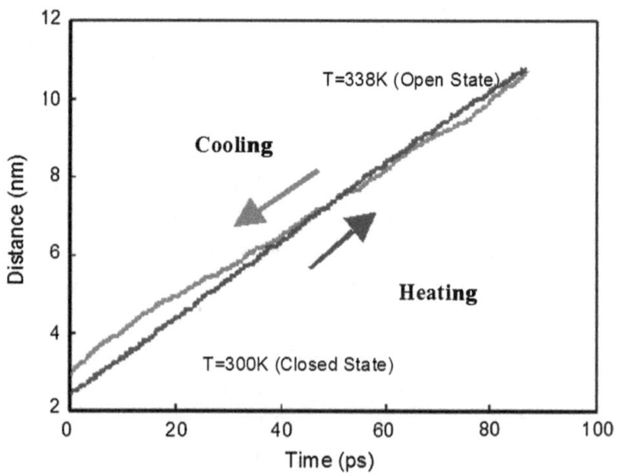

Fig. 3.42 Reversibility of the nanogripper when increasing and decreasing the temperature from $T = 300$ K to 338 K

Optimization of the Nanogripper

The optimization of the nanogripper characteristics implies the re-design of the B-DNA molecule. Two solutions are thus possible.

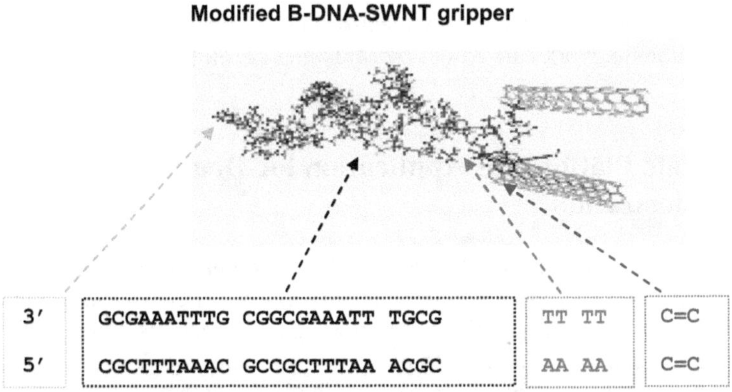

Fig. 3.43 (Color online) Optimization of modified dsDNA-SWNT nanogripper design using rich base pairs (4 AT base pairs) near the SWNT gripper

1. *DNA unzipping using rich base pairs*: The main idea is to synthesize sequence-specific base pairs with respect to their hydrogen bonds. One possible way is to synthesize a DNA base pairing where were AT base pair rich in the region near the SWNT gripper and GC rich in the region further away. Since AT base pairs are made using only two hydrogen bonds, they will be easier to break than GC pairs. It means that the more rich in AT the DNA is closer to the SWNTs, the easier it will be to open the nanogripper while retaining structural integrity in the entire device. Some simulations demonstrated that the sequence-specific base-pairing forces of DNA were on the order of 9 pN for AT ($g_0 = -11.6248$ kcal/mol) and 20 pN for GC ($g_0 = -26.17$ kcal/mol) base pairs. These values are consistent with earlier results by Rouzina and Bloomfield [73]. An example is given in Fig. 3.43 where we introduced four AT base pairs (underlined in red) rich in the region near the SWNT gripper symbolized by C=C carbon linkers attached to the 5′–3′ ends.
2. *DNA bending using purely electrostatic mechanism:* The actuation mechanism can be also improved by the creation of like electrostatic charges along each DNA strand which forces the two strands to repel each other (unzipping DNA sequence) and move apart thus creating an opening motion of the gripper. Such laterally asymmetric charge distributions should generate bending forces by local reduction (cations) or enhancement (anions) of interphosphate repulsions (at positions ∼20–24 with GC base pairs). According to this, the asymmetric reduction or enhancement of electrostatic repulsive forces on DNA faces cause DNA bending by unbalancing the repulsive forces of the two sides of the double helix. This motion can be reversed by neutralizing the charges. Creation of electrostatic charges depends on the differences in the ionization states of the base pairs which in turn depends on the pH of the solvent. Thus by the introduction of different amino acids along the DNA strands and varying the pH of the solvent different nanogripper architecture with degree of gripping motion can be obtained. Placement of fixed charges near one face of the DNA double helix in-

ducing DNA bending by a purely electrostatic mechanism is in agreement with the experimental work carried out by McDonald et al. [74].

6 Multiscale Platform as Application for Drug Delivery Characterization

The interactive simulation platform allows to conceptualize some generic scenarios of nanovectors and biomolecular interactions for efficient drug delivery. The nanovector is coated with specific recognition ligands (peptides, glycoproteins) that bounds to the cell surface in order to perform a site selective delivery. Nanovector-cell interaction occurs and insertion through the cell membrane is initiated. Recent experimental investigations in targeted drug-delivery systems [75] have shown that the fundamental mechanisms of nanovector-cell interactions and insertion in living cells are not well understood. A better understanding of the atomic level interactions is necessary in designing drug's selectivity mechanisms, cell-nanovector interaction and insertion in cell membrane. Present limitations are far from being addressed with the conventional drug carriers: the ability to penetrate the cellular membrane, to enter the cytoplasm, and to transport drug molecules to the near vicinity of the nucleus. To optimize these processes, a realistic model of the membrane and surrounding environment is needed. However, despite many recent successes, such simulations are hampered by the accessible system size and timescale. To glean some insights into possible mechanisms associated with the membrane insertion process we will use molecular dynamics calculations coupled to quantum mechanics to investigate the applicability of f-CNTs as drug delivery agents and the effects of plasma membrane structure.

Simulations in Fig. 3.44 reveals two opposite (negative and positive) electrostatic regions. Thus, a high attractive energy between f-CNT and lipids allow a possible nanovector penetration into lipids. Next, random thermal fluctuations drive one end of the f-CNT toward the core of the membrane.

7 Conclusion

Development of bionano components from biological systems is the first step towards the design and development of advanced bio-nanorobots. Design based on multiscale coupled with virtual reality advanced techniques approach is very promising in the domain of bio-nanorobotics. We have demonstrated that realistic designs and simulations can be carried out by integrating the physics at several scales (atomistic, nanometer, molecular and mesoscopic) for various time scales of simulation (from few nanoseconds to several milliseconds). Various modeling tools have been used:

- *Steered Molecular Dynamics* (*SMD*): These SMD calculations provide a complementary computational approach to the Atomic Force Microscopy and optical

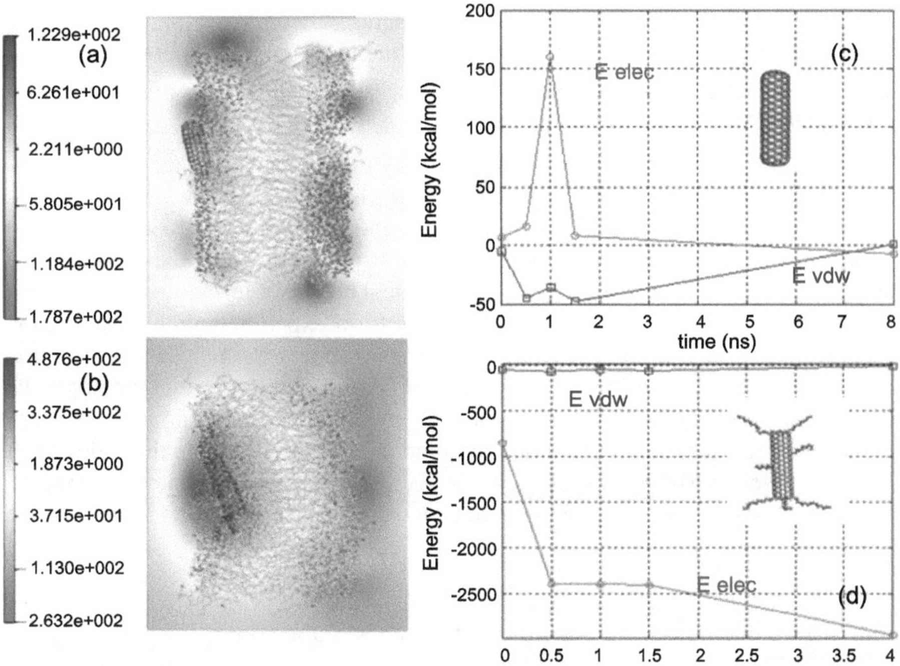

Fig. 3.44 Characterization of nanovector interaction with lipids bilayer. (**a**) The electrostatic potential map represents random distributed energies, and small attractive forces. The interaction between, not functionalized CNT/lipids are not strongly pronounced in order to demonstrate attraction. (**b**) Two opposite (negative and positive) electrostatic regions, thus a high attractive energy between f-cnt and lipids allow a possible f-cnt penetration into lipids. (**c**) Energies of interaction between not functionalized CNT and lipids. (**d**) Energies of interaction between functionalized CNT and lipids

tweezers in the study of protein-based nanosprings. As time-dependent external forces are applied to the molecular springs by pulling one end of the molecule with a cantilever spring, performance comparison of protein-based elastic joints has been rendered possible. These biosprings are classified following their mechanical characteristics: elasticity, reversibility, displacement, linearity, relaxation time and strain.

- *Quantum Mechanics, Steered Molecular Dynamics and Continuum Mechanics*: The mechanical properties of long protein-based passive kinematic chains requires simulation times around hundreds 10^6 seconds that can be reduced to very short time scale length of simulation 10^{-12} seconds using multiscale modeling. As example, dynamic characteristics of molecular sarcomere mechanism composed of proteins-based passive kinematic chains and actuated by myosin protein-based nanoactuator has been successfully simulated.
- *Quantum Mechanics and Molecular Dynamics*: Finally, nanoscale robotic devices actuated by DNA molecule have been investigated from design, modeling and simulation perspectives using Quantum mechanics and Molecular Dynamics. These tools permit the optimized design of an encapsulated DNA-double

walled carbon nanotube (DNA@MWNT) actuator from its initial design stage to its position and force control stage. We demonstrated that these characteristics are strongly dependent on medium properties: temperature, protonation and electron density control parameters.

While a more careful evaluation of the methodology is ongoing and will be continued in the future, the purpose of this chapter is to demonstrate the capabilities of the design and simulation tool set. Due to the relatively small size (<50 nm) and controlled biological environment conditions single molecule and protein experiments cannot be used to study their dynamical behavior during this Ph.D. work. Future work will be carried out in order to perform experimental characterization using Atomic Force Microscopy or laser tweezers apparatus in close collaboration with biology and nanotechnology laboratories. As an alternative methodology, we investigated in Chap. 4 novel co-prototyping methods necessitating a close cooperation between experimentation and simulation. A close collaboration with Prof. Bradley J. Nelson (IRIS-ETH Zurich, Switzerland) has been initiated in order to simulate and characterize new engineered nanostructures based on carbon nanotube technology.

References

1. Calvacanti, A., Shririnzadeh, B., Freitas, R., Hogg, T.: Nanorobot architecture for medical target identification. Nanotechnology **19**, 1–15 (2008)
2. Requicha, A.: Nanorobots, NEMS and nanoassembly. Proc. IEEE Sens. **91**(11), 1922–1933 (2003)
3. Hogg, T., Kuekes, P.J.: Mobile microscopic sensors for high resolution in vivo diagnostics. Nanomed. Nanotechnol. Biol. Med. **2**(4), 239–247 (2006)
4. Yesin, K.B., Vollmers, K., Nelson, B.J.: Modeling and control of unthered biomicrorobots in a fluidic environment using electromagnetic fields. Int. J. Rob. Res. **25**(5–6), 527–536 (2006)
5. Mathieu, J.-B., Beaudoin, G., Martel, S.: Method of propulsion of a ferromagnetic core in the cardiovascular system through magnetic gradients generated by an MRI system. IEEE Trans. Biomed. Eng. **53**(2), 292–299 (2006)
6. Behkam, B., Sitti, M.: Design methodology for biomimetic propulsion of miniature swimming robots. J. Dyn. Syst. Meas. Control **128**, 36–43 (2006)
7. Martel, S., Tremblay, C., Ngakeng, S., Langlois, G.: Controlled manipulation and actuation of microobjects with magnetotactic bacteria. Appl. Phys. Lett. **89**, 233804 (2006)
8. Sharma, G., Rege, K., Budil, D., Yarmush, M., Mavroidis, C.: Computational studies of protein based nanoGripper. Int. J. Rob. Res. (2009)
9. Steager, E., Kim, C.B., Patel, J., Bith, S., Naik, C., Reber, L., Kim, M.J.: Control of microfabricated structures powered by flagellated bacteria using phototaxis. Appl. Phys. Lett. **90**, 263901 (2006)
10. Howard, J.: Mechanics of Motor Proteins and the Cytoskeleton. Sinauer Associates, Sunderland (2001), 367 pp.
11. Montegano, C.D., Bachand, G.D.: Constructing nanomechanical devices powered by biomolecular motors. Nanotechnology **10**, 225–331 (1999)
12. Amos, W.B., et al.: Calcium binding proteins in a Vorticellid contractile organelle. J. Cell Sci. **190**, 203–213 (1975)
13. Yu, H.-h., Swager, T.M.: Molecular actuators—designing actuating materials at the molecular level. IEEE J. Ocean. Eng. **29**(3), 692–695 (2004)

14. Huang, T.J., Flood, A.H., Brough, B., Liu, Y., Bonvallet, P.A., Kang, S., Chu, C.-W., Guo, T.-F., Lu, W., Yang, Y., Stoddart, J.F., Ho, C.-M.: Understanding and harnessing biomimetic molecular machines for NEMS actuation materials. IEEE Trans. Autom. Sci. Eng. **3**(3), 254–259 (2006)
15. Sauvage, J.-P., Dietrich-Buchecker, C.: Molecular Catenanes, Rotaxanes and Knots. VCH, Weinheim (1999)
16. Huang, T.J., et al.: A nanomechanical device based on linear molecular motors. Appl. Phys. Lett. **85**, 5391–5393 (2004)
17. Mahadevan, L., Matsudaira, P.: Mobility powered by supramolecular springs and ratchets. Science **288**, 95–99 (2000)
18. Kitamura, K., Tokunaga, M., Iwane, A.H., Yanagida, Y.: A single myosin head moves along an actin filament with rectangular steps of 5.3 nm. Nature **397**, 129 (1999)
19. Shingyoji, C., Higuchi, H., Yoshimura, M., Katayama, E., Yanagida, T.: Dynein arms are oscillating force generators. Nature **393**, 711–714 (1998)
20. Block, S.M.: Kinesin what gives? Cell **93**, 5–8 (1998)
21. Knoblauch, M., Naull, G.A., Muller, T., Prufer, D., Schneider-Huther, I., et al.: ATP-independent contractile proteins from plants. Nat. Mater. **2**, 600–603 (2003)
22. Atsumi, T., McCarter, L., Ymae, Y.: Polar and lateral flagellar motors of marine Vibrio are driven by different ion-motive forces. Nature **355**(6356), 182–184 (1992)
23. Li, J., Tan, W.: A single DNA molecule nanomotor. Nano Lett. **2**, 315–318 (2002)
24. Jorgenson, W.L., Chandrasekhar, J., Madura, D., Impey, R.W.: Comparison of simple potential functions for simulating liquid water. J. Chem. Phys. **79**, 926–935 (1993)
25. Isralewitz, B., Baudry, J., Gullingsrud, J., Kosztin, D., Schulten, K.: Steered molecular dynamics investigations of protein function. J. Mol. Graph. Model **19**(12), 13–25 (2001)
26. Park, S., Khalili-Araghi, F., Tajkhorshid, E., Schulten, K.: Free energy calculation from steered molecular dynamics simulations using Jarzynski's equality. J. Chem. Phys. **119**(6), 3559–3566 (2003)
27. Bustamante, C., Smith, S.B., Liphardt, J., Smith, D.: Single-molecule studies of DNA mechanics. Struct. Biol. **10**, 279–285 (2000)
28. Strick, T.R., Charvin, G., Dekker, N.H., Allemand, J.-F., Bensimon, D., Croquette, V.: Tracking enzymaic steps of DNA topoisomerases using single-molecule micromanipulation. C. R. Phys. **3**, 595–618 (2002)
29. Williams, M.C., Rouzina, I.: Force spectroscopy of single DNA and RNA molecules. Curr. Opin. Struct. Biol. **12**, 330–336 (2002)
30. Sattin, B.D., Pelling, A.E., Goh, M.C.: DNA base pair resolution by single molecule force spectroscopy. Nucleic Acids Res. **32**(16), 4876–4883 (2004)
31. Essevaz-Roulet, B., Bockelmann, U., Heslot, F.: Mechanical separation of the complementary strands of DNA. Proc. Natl. Acad. Sci. USA **94**, 11935–11940 (1997)
32. Cocco, S., Monasson, R., Marko, J.F.: Force and kinetic barriers to unzipping of the DNA double helix. Proc. Natl. Acad. Sci. USA **98**(1), 8608–8613 (2001)
33. Strom, C., Nelson, P.C.: Theory of high-force DNA stretching and overstretching. Phys. Rev. **67**, 051906 (2003)
34. Noy, A., Perez, A., Lankas, F., Luque, F.J., Orozco, M.: Relative flexibility of DNA nad RNA: a molecular dynamics study. J. Mol. Biol. **343**, 627–638 (2004)
35. Granzier, H., Labeit, S.: Cardiac titin: an adjustable multi-functional spring. J. Phys. **541**, 335–342 (2002)
36. Lu, H., Schulten, K.: Steered molecular dynamics simulations of force-induced protein domain unfolding. Proteins **35**, 453–463 (1999)
37. Gao, M., Graig, D., Vogel, V., Schulten, K.: Identifying unfolding intermediates of FN-III10 by steered molecular dynamics. Matrix Biol. **232**, 939–950 (2002)
38. Paci, E., Karplus, M.: Forced unfolding of fibronectin type 3 modules: an analysis by biased molecular dynamics simulation. J. Mol. Biol. **288**, 441–459 (1999)
39. Marszalek, P.E., Lu, H., Li, H., Carrion-Vazquez, M., Oberhauser, A.F., Schulten, K., Fernandez, J.M.: Mechanical unfolding intermediates in titin modules. Nature **402**, 100–103 (1999)

40. Vazina, A.A., Lanina, N.F., Alexeev, D.G., Bras, W., Dolbnya, I.P.: The structural principles of multidomain organization of the giant polypeptide chain of the muscle titin protein: SAXS/WAXS studies during the stretching of oriented titin fibres. J. Cell Biol. **155**(2), 251–262 (2006)
41. Schoenauer, R., Bertoncini, P., Machaidze, G., Aebi, U., Perriard, J.-C., Hegner, M., Agarkova, I.: Myomesin is a molecular spring with adaptable elasticity. J. Mol. Biol. **349**, 367–379 (2005)
42. Gao, M., Lu, H., Schulten, K.: Simulated refolding of stretched titin immunoglobulin domain. Biophys. J. **8**, 2268–2277 (2006)
43. Lu, H., Schulten, K.: The key event in force-induced unfolding of titin. Biophys. J. **79**, 51–65 (2000)
44. Cramer, C.J.: Essentials of Computational Chemistry, pp. 191–232. Wiley, New York (2002)
45. Stewart, J.J.P.: Optimization of parameters for semiempirical methods IV: extension of MNDO, AM1, and PM3 to more main group elements. J. Mol. Model. **10**, 155–164 (2004)
46. Kuhn, B., Kollman, P.A.: Binding of a diverse set of ligands to avidin and streptavidin: an accurate quantitative prediction of their relative affinities by a combination of molecular mechanics and continuum solvent models. J. Med. Chem. **43**, 3786–3791 (2000)
47. Liu, Y.J., Chen, X.L.: Continuum models of carbon nanotube-based composites using the boundary element method. J. Bound. Elem. **1**(2), 316–335 (2003)
48. Gao, M., Wilmanns, M.: Steered molecular dynamics study of titin I1 domain unfolding. Biophys. J. **83**, 3435–3445 (2002)
49. Trombitas, K., Gasser, M., Labeit, S., Jin, J.-P., Kellermayer, M., Helmes, M., Granzier, H.: Titin extensibility in situ: entropic elasticity of permanently folded and permanently unfolded molecular segments. J. Cell Biol. **140**, 853–859 (1998)
50. Zhang, B., Guangzhao, X., Evans, J.S.: A kinetic molecular model of the reversible unfolding and refolding of Titin under force extension. Biophys. J. **77**, 1306–1315 (1999)
51. Day, R., Bennion, B.J., Daggett, V.: Increasing temperature accelerates protein unfolding without changing the pathway of unfolding. J. Mol. Biol. **322**, 189–203 (2002)
52. Minajeva, A., Kulke, M., Fernandez, J.M., Linke, W.A.: Unfolding of titin domains explains the viscoelastic behavior of skeletal myofibrils. J. Struct. Biol. **80**, 1442–1451 (2001)
53. Dubey, A., Sharma, G., Mavroidis, C., Tomassone, M., Nitkitczuk, K.P., Yarmush, M.L.: Computational studies of viral protein nano-actuators. J. Comput. Theor. Nanosci. **1**, 18–28 (2004)
54. Sherman, W.B., Seeman, N.C.: A precisely controlled DNA biped walking device. Nano Lett. **4**(7), 1203–1207 (2004)
55. Simmel, F.C., Yurke, B.: Using DNA to construct and power a nanoactuator. Phys. Rev. E **63**, 127–132 (2001)
56. Yan, H., Zhang, X., Shen, Z., Seeman, N.C.: A robust DNA mechanical device controlled by hybridization topology. Nature **415**, 62–65 (2002)
57. Hummer, G., Rasaiah, J.-C., Noworyta, J.P.: Water conduction through the hydrophobic channel of a carbon nanotube. Nature **414**, 188–190 (2001)
58. Joseph, S., Mashl, R.J., Jakobsson, E., Aluru, N.R.: Electrolytic transport in modified carbon nanotubes. Nano Lett. **3**, 1399–1403 (2003)
59. Wei, C., Srivastava, D.: Theory of transport of long polymer molecules through carbon nanotube channel. Phys. Rev. Lett. **91**, 235901 (2002)
60. Zheng, Q., Jiang, Q.: Multiwalled carbon nanotubes as gigahertz oscillators. Phys. Rev. Lett. **88**, 045503 (2002)
61. Fennimore, A.M., et al.: Rotational actuators based on carbon nanotubes. Nature **424**, 408–410 (2003)
62. Dong, L., Nelson, B.J., Fukuda, T., Arai, F.: Towards nanotube linear servomotors. IEEE Trans. Autom. Sci. Eng. **3**(23), 228–235 (2006)
63. Okada, T., Kaneko, T., Hatakeyama, R.: Single-stranded DNA insertion into single-walled carbon nanotubes by ion irradiation in an electrolyte plasma. Jpn. Soc. Appl. Phys. **45**, 8335–8339 (2006)
64. Gao, H., Kong, Y., Cui, D., Ozken, G.S.: Spontaneous insertion of DNA oligonucleotides into carbon nanotubes. Nano Lett. **3**(4), 471–473 (2003)

65. Steger, G.: Thermal denaturation of double-stranded nucleic acids: prediction of temperatures critical for gradient gel electrophoresis and polymerase chain reaction. Nucleic Acids Res. **22**(14), 2760–2768 (1994)
66. Williams, M.C., Wenner, J.R., Rouzina, I., Bloomfield, V.A.: Entropy and heat capacity of DNA melting from temperature dependence of single molecule stretching. Biophys. J. **80**, 1932–1939 (2001)
67. Williams, M.C., Wenner, J.R., Rouzina, I., Bloomfield, V.A.: Effect of pH on the overstretching transition of double-stranded DNA: evidence of force-induced DNA melting. Biophys. J. **80**, 874–881 (2001)
68. Balzani, V., Venturi, M., Credi, A.: Molecular Devices and Machines: A Journey into the Nanoworld. Wiley–VCH, New York–Weinheim (2003)
69. Wenner, J.R., Williams, M.C., Rouzina, I., Bloomfield, V.A.: Salt dependence of the elasticity and overstretching transition of single DNA molecules. Biophys. J. **82**, 3160–3169 (2002)
70. Punkkinen, O., Hansen, P.L., Miao, L., Vattulainen, I.: DNA overstretching transition: ionic strength effects. Biophys. J. **89**, 967–978 (2005)
71. He, P., Li, S., Dai, L.: DNA-modified carbon nanotubes for self-assembling and biosensing applications. Synthetic Mater. **154**, 17–20 (2005)
72. Kohandel, M., Ha, B.-Y.: Thermal denaturation of double-stranded DNA: effect of base stacking. Phys. Rev. E **73**, 011905 (2006)
73. Rouzina, I., Bloomfield, V.A.: Force-induced melting of the DNA double helix 1. Thermodynamic analysis. Biophys. J. **80**, 882–893 (2001)
74. McDonald, R.J., Dragan, A.I., Kirk, W.R., Neff, K.L., Privalov, P.L., Maher, L.J.: DNA bending by charged peptides: electrophoretic and spectroscopic analyses. Biochemistry **46**, 2306–2316 (2007)
75. Bianco, A., Kostarelos, K., Prato, M.: Applications of carbon nanotubes in drug delivery. Curr. Opin. Chem. Biol. **9**, 674–679 (2005)

Chapter 4
Characterization and Prototyping of Nanostructures

1 Introduction

Prototyping is an iterative process that is part of the analysis phase of the nanosystems development life cycle. Usually, technologists expect the performance of the ultimate nanosystem to be the same as the designed prototype. However, the interactive multiscale prototyping presented in Chap. 3 leads to insufficient analysis at the pre-project stage, degraded performances of the ultimate nanodevice due to experimentation limitations, and lack in quantifiable experimentation feedback. Taking into account the disadvantages of the previous method, we present in this chapter a novel methodology of design and optimization using co-prototyping concepts of carbon nanotube-based nanodevices. It is an iterative optimization process coupling experiments performed on real nanodevices with computational methods of simulation. Among its' many benefits, the co-prototyping approach lowers the development cost, the development time involved in prototyping, allows for more iterations during experimentation, and gives developers the chance to get immediate experimental feedback on refinements to the design. Furthermore, the theoretical characterization allows to analyze in deep detail some novel properties that could not be observable by the used experimental platform.

As illustration of the present co-prototyping concept, we studied nanoscale structures using multiple, but independent, bi-directional linear bearings constructed within individual multiwalled nanotubes. In the first section, we present the behavior description of the 'head-to-head' carbon nanotube-based nanodevice. The dedicated methods used for modeling, simulation and optimization are then presented. The main outcomes are the explanation of the system behavior, the optimization of the gap distance and the number of nanotube shells with respect to the electric driving voltage. The second section presents a novel rotational nanodevice that has been designed based on theoretical observations.

This work has been developed in close collaboration with Prof. Bradley J. Nelson and Dr. Lixin Dong from the Institute of Robotics and Intelligent Systems at ETH Zurich in Switzerland. The results presented in this chapter have been initiated during my stay at ETHZ during the period (May 2008–September 2008) and pursued until now.

M. Hamdi, A. Ferreira, *Design, Modeling and Characterization of Bio-Nanorobotic Systems*,
DOI 10.1007/978-90-481-3180-8_4,

2 Characterization of NEMS Based on Linear Bearings

Supermolecular bearings based on inter-shell displacements in multiwalled nanotubes (MWNTs) have been demonstrated to be ultra-fast, fatigue free constructs with very low friction [1, 2]. Apart from their favourable nanomechanical performance, the variation of carbon nanotube resistance with telescoping core motion [4] provides a unique mechanism to electrically sense and control their operation. These electromechanical building blocks can be used to create tubular switches [5, 7, 10], memories [8], nano-servomotors with integrated position sensing [6], and rotational elements for use in Nano Electro Mechanics System (NEMS) [15–17] location, length and shells in order to create structures where some shells can axially translate or rotate with respect to others [10].

The design and prototyping of complex systems comprising hardware and software elements is of considerable interest on account of its extremely varied applications. It is of fundamental importance to optimize both the cost and the performance of systems based on inter-shell displacements in MWNTs. Co-prototyping is an approach to the development of systems composed by both hardware and software modules. Its main purpose is to optimize the trade-off between experiment performance and software reconfigurability. Moreover the aim of co-prototyping is to be able to prototype a whole system without excessive preliminary constraints on mapping the module onto hardware and software parts.

In this work we propose a co-prototyping methodology which is not only suitable for the above mentioned systems, but can also be extended to more complex telescoped nanodevice based carbon nanotube. The designed systems to which our co-prototyping methodology is applied are electromechanical switches based on bi-directional linear bearings. Our proposed architecture template is depicted in Fig. 4.1. Co-prototyping mainly consists on coupling experiments to theoretical characterization.

In order to reduce the development cost and time during the nanodevice prototyping, computational methods of simulation are used to palliate the lack of measurement data.

It is not possible to measure and observe certain characterizations, such as the determination by SEM of number of shells in a structure MWNT or precise measurement of the GAP distance in head to head nanotubes shuttles structure (Fig. 4.7). Simulations by molecular dynamics enable us to determine these parameters and their evolution according to the state of system.

We co-prototyped the head to head nanotubes device (Fig. 4.7). the method is subdivided in 6 phases, each phase corresponds to a well defined task of prototyping. The Co-prototyping task starts with the specification of device model. From the platform experimental, phase 1, we gather dimensions of the nanodevice. In phase 2, because we work in the molecular scale, we generate the 3D the molecular model, which consists in creating the coordinates (xyz) atoms. Then we use molecular mechanics to find the geometry of the molecule (or an assemblage of molecules) which corresponds to a minimum potential energy. The nanodevice is actuated by an electric potential, we take in consideration in the model the atomic electrostatic interactions. The electrostatic charge distribution on the carbon nanotube is obtained by an

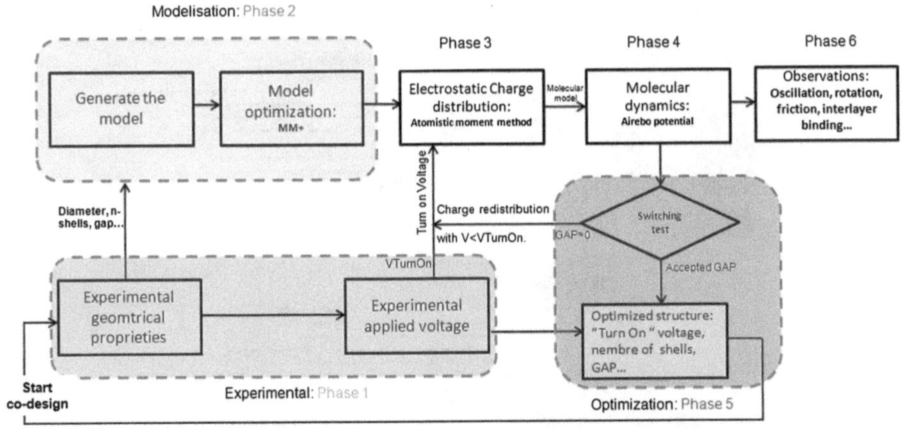

Fig. 4.1 Co-prototyping method for based carbon nanotube based nanostructures. The method contains 6 phases: The phases 2 and 3 communicate with the experimental phase 1 to optimize and update the model of the system to be prototyped. Phase 4 and 6 allow characterization and optimization of nanodevice behavior, phase 6 allow to observe certain phenomenons that are not observable by experimentations

atomistic moment method based on classical electrostatics theory nanotubes. In the phase 4, we study the dynamics of the system. We simulate the nanodevice behavior by using molecular dynamics. Classical molecular dynamics with AIREBO potential [3] are employed in the simulation. The potential used for this work is based on the reactive empirical bond-order potential of Brenner. In phase 5 simulations allow us to optimize some parameters of the prototype, such as calculate the electric tension corresponding to different gap distance or to get the best number of shells for each potential applied to actuate the device. In phase 6, realistic simulations enable us to observe certain phenomena which cannot be observed by our experimental platform, such as the interlayer friction, structure rotation, oscillation or nanotube defect by high energy. in the paragraphs following, we will present the results and the discussions of each phase.

Nanostructure and Device Overview

The structure presented here employs multiple, but independent, bi-directional linear bearings constructed within individual MWNTs it is built using the piecewise shell engineering technique. A NEMS switch based on the inter-shell displacement mechanisms within these structures is demonstrated using experimental results. Computational models provide new insights into the structural and energetic stability aspects of these unique nanomechanical elements. Based on the modeling results, it is anticipated that these elements can serve as novel, re-configurable NEMS building blocks with actively tunable structural stiffness/operating frequency. Another

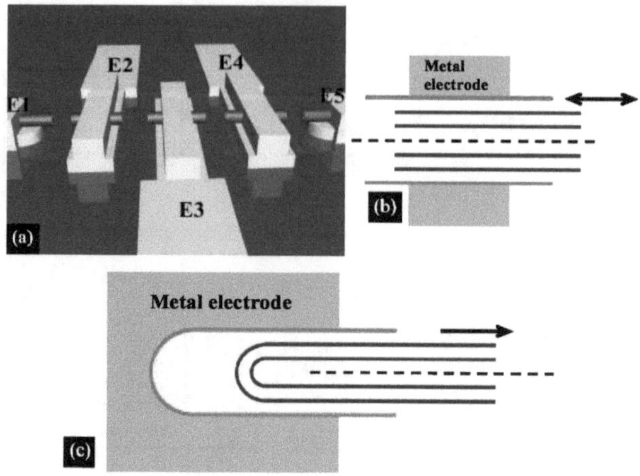

Fig. 4.2 (Color online) Nanostructure overview. (**a**) A piecewise engineered MWNT bridging five spatially separated electrodes. The NT is further engineered to create gaps in each of the suspended, inter-electrode segments. The electrode terminology ('E1' through 'E5') that would be used in the discussions within this chapter are highlighted here. (**b**) The shell architecture at electrodes 'E2', 'E3' and 'E4'. It is important to note that the outermost shell, which remains anchored to the electrodes, is shown in *red* while the inner shells that can displace axially are shown in *green*. The NT axis is shown in *black dotted lines*. (**c**) The shell architecture at the distal electrodes 'E1' and 'E5' showing the unidirectional nature of bearings at these locations

potential application that has been identified using appropriate models relates to microwave resonators enabled by the van der Waals interactions within this inter-shell displacement mode. The key components of the nanostructure, which is schematically illustrated in Fig. 4.2(a), include:

1. A MWNT assembled onto five spatially separated electrodes using the floating electrode Dielectrophoresis technique. The electrodes are designed with a 220 nm pitch.
2. Piecewise shell engineering of the NT to create breaks in each of the four nanotube (NT) segments that are suspended in air between the electrodes. After current induced shell etching, the inter-segment gaps are found to be in the 6 to 15 nm range.
3. Each of the three inner electrodes (labeled E2 through E4 in Fig. 4.2(a)) has an open, cap-less NT segment that is anchored at the outermost shell. The inner shells of the NT are free to slide axially within the outer housing in either direction. The cross-sectional schematic of these NT segments is shown in Fig. 4.2(b), and clearly reveals the unique shell architecture.
4. At the distal electrodes (labeled E1 and E5 in Fig. 4.2(a)), the NT segment is anchored at the outermost shell and the inner shells are free to slide within this outer housing. However, the NT caps and the metal electrode constrain the axial displacement to occur in only one direction within these segments. This is illustrated in panel (c) of Fig. 4.2.

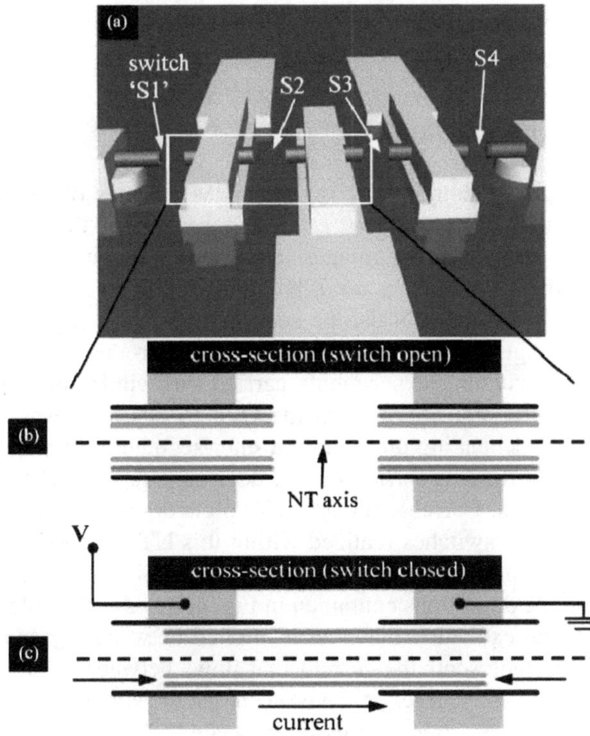

Fig. 4.3 High-density electromechanical switches based on bi-directional linear bearing. (**a**) Schematic illustration of four switches constructed within a MWNT. The labeling of switches from 'S1' through 'S4' is also shown here. (**b**) Cross-sectional view of one of these switches in the 'open' state. (**c**) Cross-sectional view of a switch in the 'closed' state when an excitation bias is applied

One possible application of the nanostructure relates to electromechanical nanoswitches based on axial shell displacements. The co-prototyped nanodevice is schematically illustrated in Fig. 4.3. Each pair of spatially separated, opposing nanotube segments located between adjacent metal electrodes constitute a switching device, and these are labeled S1 through S4 as shown in panel (a). In the as-fabricated configuration, all four switches are in the OFF state (Fig. 4.3(b)). When a potential difference is applied between any two adjacent metallic contacts, charges with opposing polarities are induced in the nanotube segments that are attached to these contacts. This charge distribution results in two types of electrostatic forces: (1) an intra-segment repulsive force (due to like charges) that extrudes the core from the outer housing, and (2) an inter-segment attractive force (due to charges with opposing polarities) that causes the core shells within each NT segment to move towards each other. When the applied excitation bias exceeds a characteristic threshold value, the combined contributions of these two electrostatic force components would be higher than the van der Waals interactions within each of the two segments, and result in a closure of the gap Fig. 4.3(c). In this ON state of the switch, a current flow is established between the metallic contacts.

Characterization of the Shuttle Nanostructure Switching: Phase 4 and 5

Experimental Results

Scanning Electron Microscopy (SEM) images of fabricated NEMS switches are shown in Figs. 4.4(a)–(c), while a section of the nanoarray design is shown in panel (d). From the images taken with a stage tilt (panels (b)–(c)), it is evident that the NT segments are fully suspended in air in the region between metallic contacts. In addition, it can be seen that the shell-engineered gaps are located at the mid-lengths of NT segments between electrodes.

In switching experiments carried out within this effort, the threshold bias required for transitions from the OFF to ON state was found to vary between 0.8 to 6 V, depending on the initial inter-segment separation. The results from switching experiments carried out on devices constructed at one NT-site are highlighted in Fig. 4.5. Current-voltage (I–V) plots demonstrating the operation of four supermolecular switches realized within this NT are shown in panel (a). It is important to note that the terminology used for labeling switches in this image is consistent with the definitions contained in Fig. 4.4(a). In these plots, the x-axis represents the external excitation bias that is applied between adjacent metallic contacts and the y-axis represents the current that flows between these respective biasing contacts. From the I–V curves, it can be seen that the voltages required for establishing an ON state in these four devices lie between 0.8 to 2 V. This variation in threshold bias can be attributed to the differences in inter-segment gaps at each of the four devices. SEM imaging during device characterization provides direct evidence for inter-shell displacements induced by electrostatic actuation. Figure 4.5 shows a switch in the OFF state with the gap clearly visible, while panel (c) shows the gap closed due to electromechanical actuation of the core shells resulting in an ON state.

Also, the increase in current beyond the threshold voltage, without a steep step, is characteristic of tunneling transport at the inter-segment interface and indicates

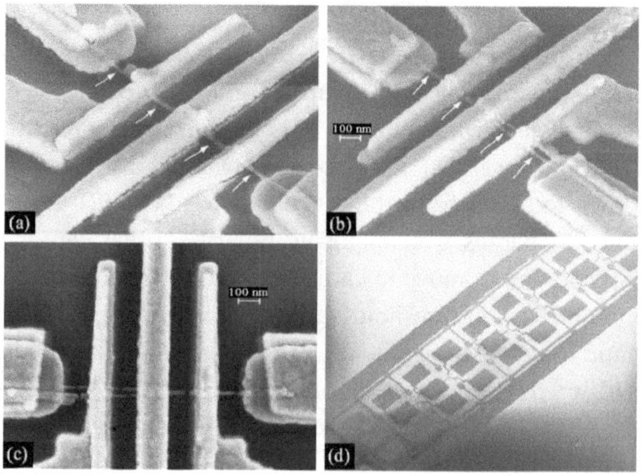

Fig. 4.4 SEM images of fabricated nanostructures, (**a**) An image taken with a stage tilt of 40 degrees. *The arrows* point to 6–15 nm gaps engineered at each of the suspended, inter-electrode NT segments. (**b**) An image of two NTs assembled at one electrode location. (**c**) Top-view of NTs in panel (**b**). (**d**) Array design

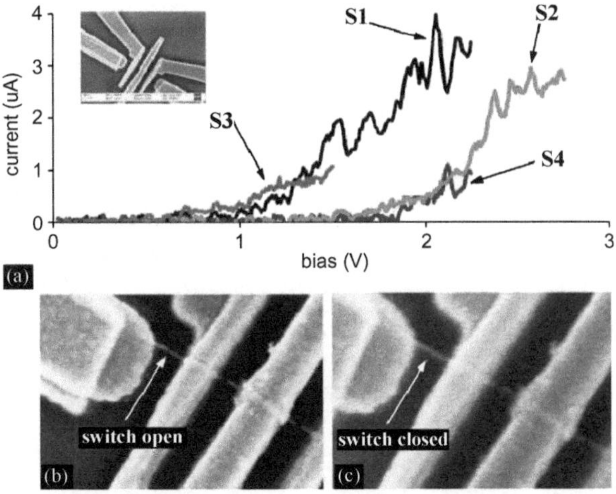

Fig. 4.5 Switching results. (**a**) I–V plots of switches with low turn-on bias (0.8 to 2 V in this case). The engineered NT is shown in *inset*. (**b**)–(**c**) SEM image of a switch in the open and closed configurations

that the carbon atoms on opposing segments do not re-establish the covalent bonds in the ON state. This behavior is different from the previous report on telescoping mechanism based NEMS switches where the NT segments remained welded to each other, most likely through covalent bonds, after switching. In this device reported by Deshpande et al., the I–V characteristic of devices exhibited a step-like increase at the threshold bias and followed a linear profile within the ON state. In switches characterized as a part of this thesis, such a behavior occurred in only one device (Fig. 4.6(c)) and the rest exhibited an I–V profile that represents tunneling transport at the interface between NT segments during the ON state [19].

Results Analysis

Explain the behavior and characteristic of MWNTs based nanodevice, require determination of precise device's characteristics such as potential energy governing the system, mechanical and electric proprieties. By experimentation it is not possible get a precise measurement this nanodevice proprieties. The solution is to characterize theoretically the nanodevice by using realistic modeling and simulation. We developed a co-prototyping method, this method allows as to calculate the potential energies governing the device and to get make some mechanical calculation such as the nanotube trajectory and delivered mechanical force's. Calculating the electric and mechanical proprieties allows as to optimize the prototyped nanodevice.

In our simulation we use AIREBO potential coupled with electrostatic forces. The electrostatic charge distribution along carbon nanotube was calculated by atomistic method moment. We make reference to *phase 4* (MD simulations) and *phase 5* (optimization) of the co-prototyping method of Fig. 4.1.

The molecular dynamics simulation has been controlled through three parameters: The turn-on voltage, the gap distance and number of nanotube shells. The simulation results (Fig. 4.8(a)) shows the intercation energies between two opposite

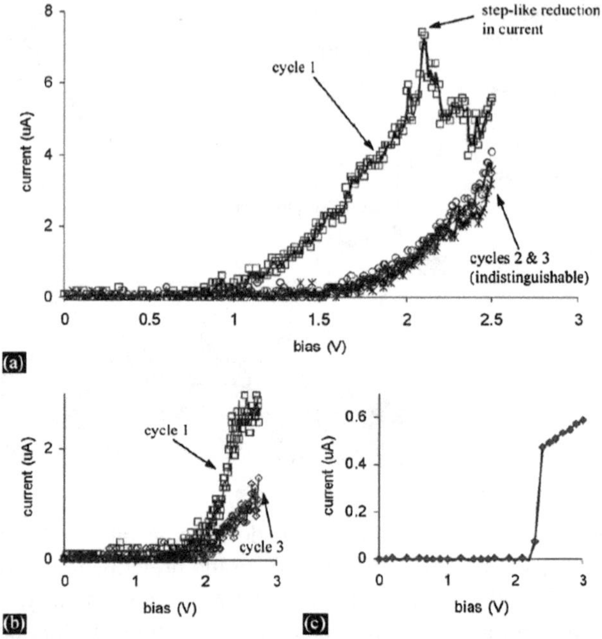

Fig. 4.6 Evolution of I–V curves during three successive switching cycles. The reduction in currents from the first to third cycle can be seen which can be attributed to a loss of carbon atoms at the contact interface in the 'ON' state. A step-like reduction in current in the first cycle (indicated with an *arrow* in image) also supports this argument. In addition, the increase in 'ON' voltage from the first to second cycle indicates an increase in device gap (due to loss of carbon atoms). (**b**) The decrease in current from the first to third cycle in another device. In panels (**a**) and (**b**) the raw data points are superimposed on a moving average curve fit (with a period of 2). (**c**) The only that exhibited a step-like increase in current

nanotubes with respect to voltage and gap distance. It should be noticed that the energy increases when the electric potential increase for a critical gap distance of 5 nm (closed state). At the same the electrostatic energy between both 'inner and outer' nanotubes (see Fig. 4.8(b)) decrease according to the applied voltage and of the gap distance. As consequence, we can say that repulsive electrostatic energy push the inner nanotubes and the attractive electrostatic energy between both opposite nanotubes allows their attraction in a closed configuration.

To optimize the structure, we calculate the number of nanotubes shells corresponding to the minimal applied electric potential. The set of the curve Fig. 4.9(a) show the evolution of the Turn-on voltage as function as the number of shells. The results show clearly the linear decrease of the turn-on voltage when the number of shells is close to 5. Then, we tried to determine the best GAP distances corresponding to a minimal applied potential as shown in Fig. 4.9(b). The minimal turn-on-voltage (1 volt) corresponds to a minimal GAP distance of 3 nm.

We studied the mechanical properties of the prototyped nanodevice. The CNT's trajectories are shown in Fig. 4.10(a), where the GAP distance is 10 mn; each CNT

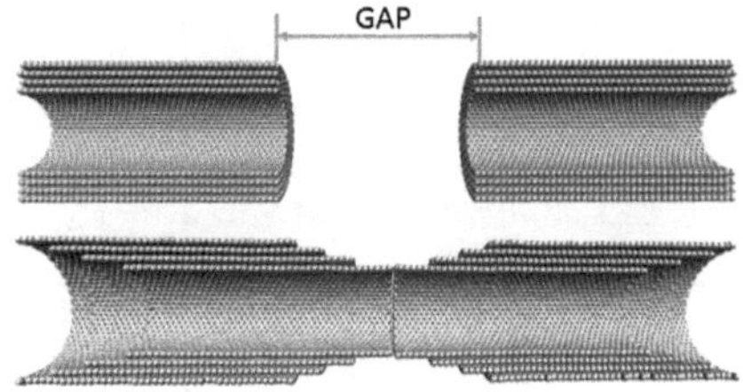

Fig. 4.7 Cross-section view of the shuttle nanostructure in On-state and Off-state. The nanostructure is composed by two head to head multiwalled nanotubes, each MWNT is composed of four shells

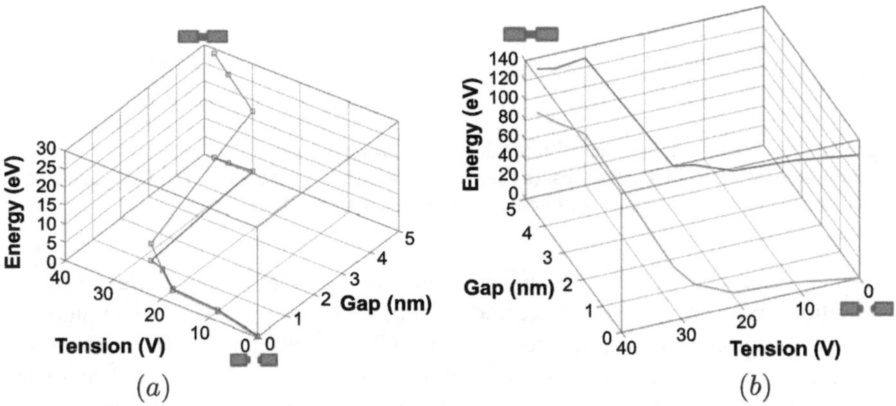

Fig. 4.8 (Color online) (**a**) Interaction energies between two opposite nanotube. *Red curve*: attractive electrostatic energy. (**b**) Interaction energies between inner and outer nanotube. *Red curve*: attractive electrostatic energy. *Blue curve*: van der Waals energy. *Blue curve*: van der Waals energy

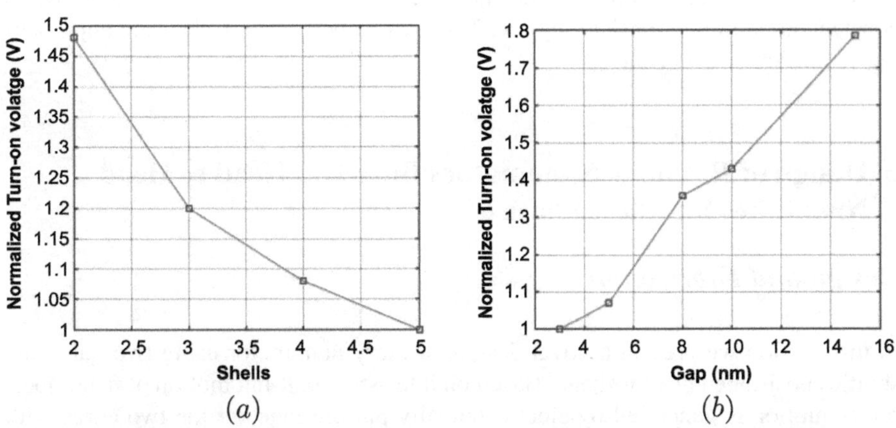

Fig. 4.9 (**a**) The figure shows the relationship between the Turn-on voltage and the number of carbon nanotube shells. (**b**) The simulation of the Turn-on voltage and gap relationship. The Turn-on voltage increase linearly and the slope of a Turn-on voltage versus gap distance plot will be related to

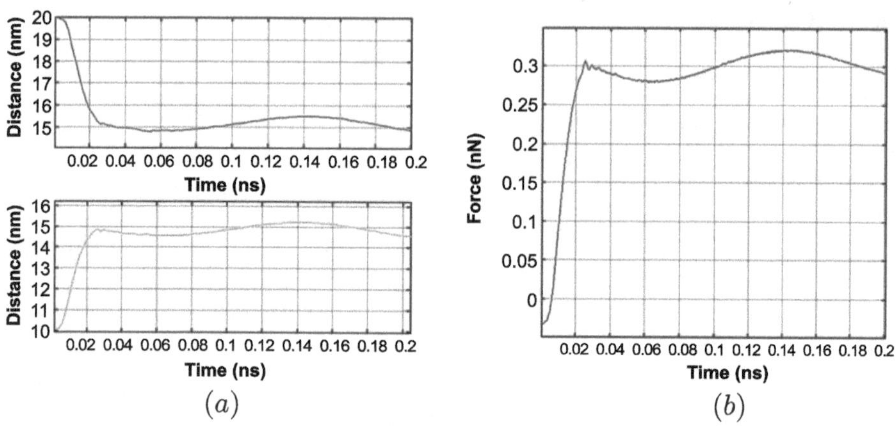

Fig. 4.10 (**a**) Inner CNT mechanical force delivered during Turn-on. The force is measured by a modified steered molecular dynamics. (**b**) Inner cnt's terminus trajectories. The GAP distance is 10 nm, the curve show an important stability of structure after Turn-on closing

move 5 nm before switching. The mechanical delivered force obtained directly from simulation. We use a modified steered molecular dynamics (SMD) technique to measure the motive force produced by inner CNT displacement. To perform this measure we ran a molecular dynamics simulation of the nanodevice with an external constraint to the inner CNT terminus. An harmonic spring of known stiffness k, has been attached to the terminus atom. The harmonic guiding potential and the corresponding exerted force are defined by:

$$U = -k(x - x_0)^2/2, \qquad F = k(x - x_0).$$

As shown in Fig. 4.10(b), the mechanical delivered force varies linearly with a constant slope in it's transient regime. Then, when both inner CNT's are approaching saturation effect occur.

3 Design of Rotatory Nanomotors Based on Head to Head Nanotubes Shuttles: Phase 6

Design and Fabrication

In this section we present a novel designed rotary nanomotor using two nanotube shuttles positioned head-to-head. Based on inter-shell bolt-nut motion of nanotubes, rotary motion is generated by electrostatically pulling together the two cores with opposite chiralities. Simulations using molecular dynamics show the generation of rotation from armchair nanotube bolt-nut pairs and their actuation properties. The ultra-compact dimensions compared to previous designs and the progress on batch fabrication of similar nanostructures promise these motors as potential building

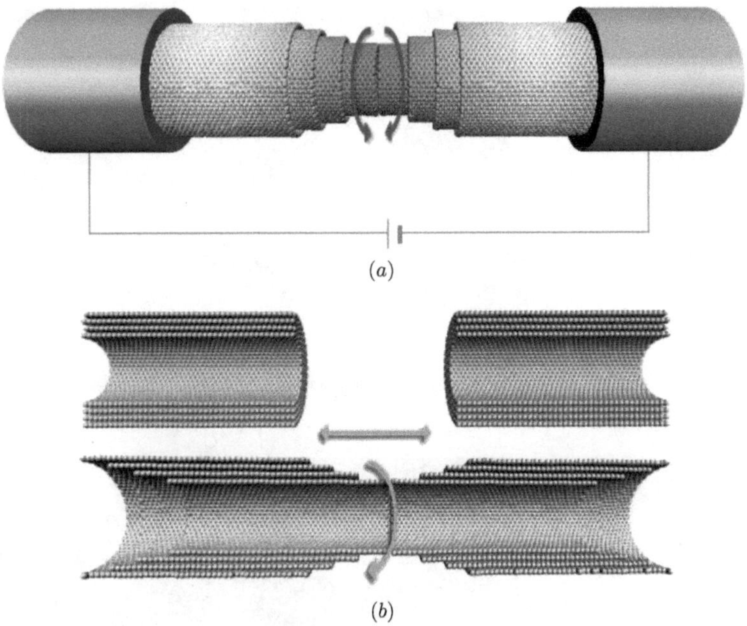

Fig. 4.11 (**a**) Schematic design of a rotary nanomotor based on head-to-head nanotube shuttles. (**b**) Cross-section view of the rotary nanomotor

blocks of nanoelectromechanical systems (NEMS) and nanorobotic systems for the applications in sensing, actuation, and computation.

The rotary nanomotor is schematically illustrated in Figs. 4.11(a) and (b). The motor consists of two MWNT segments. The outermost shells are fixed on electrodes, whereas the inner shells are freely suspended inside outer ones. By applying a DC actuation voltage, the head-to-head positioned inner tubes will slide towards each other. Due to their opposite chiralities, the inner tubes will approach to each other and both sliding and rotation, i.e., bolt-nut screwing, will be generated simultaneously. After a contact is made, the sliding motion will be constrained, whereas the rotation will remain as the only possible degree-of-freedom of motion. Important progresses have been made on the fabrication of these motors though challenges are still remaining due to the difficulty on selective synthesis of CNTs with specific chiralities. Based on CNTs with known chirality, it is possible to place two nanotubes with opposite chiralities head to head using nanorobotic manipulation [18]. Armchair nanotubes are special because their helical angle can be regarded as either $\pi/6$ or $-\pi/6$. For a combination of a pair of armchair nanotubes, the motor can be fabricated by shell engineering an individual nanotube and making it into piecewise (Figs. 4.12(a) and (b)) [19] provided the nanotubes have uniform chirality. The rotational direction in the case as using a pair of armchair nanotubes cannot be predetermined in theory, but it will continue the unidirectional rotation once the motor starts to rotate to one direction due to a random dynamic factor. Without losing generality, the molecular dynamics modeling and the working principle of the motor

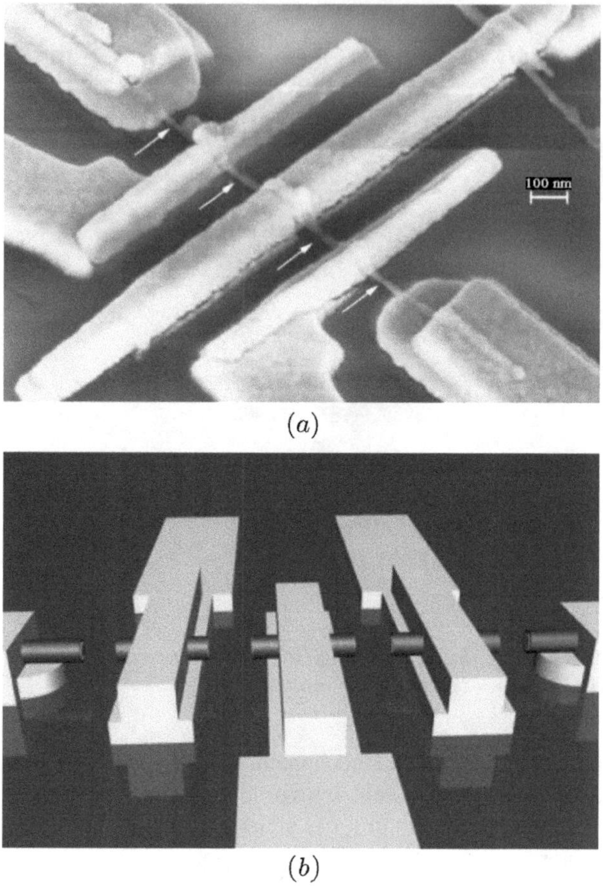

Fig. 4.12 (**a**) Shell engineered nanotube shuttles formed with a 220 nm pitch and separated by 6–10 nm gaps. *The arrows* point to the inter-segment gaps in this image taken with a stage tilt of 40°. (**b**) Schematic illustration of the core-shell mechanisms formed in panel (**a**) with the inter-segment gaps exaggerated to reveal the shell structure

are investigated in this paper by taking a rotary motor consisting of two armchair nanotubes as an example.

Molecular Dynamics Simulations on Rotatory Motor: Phase 4

Since the nanotube wall has a helical symmetry, it was recently proposed that a double walled nanotube (DWNT) can serve as a bolt-and-nut pair. Such bolt-and-nut pairs can operate, for example, as an auger of a perforating nanodrill or a nanodevice in which a force or linear motion along the nanotube axis can be transformed into a torque or rotary motion of the core tube [20]. Previously, the classification scheme for nonchiral DWNTs has been developed [21], energetic barriers to the relative sliding and rotation of walls in DWNTs [22] and to the rotation of shells in double-shell nanoparticles [23, 24] have been calculated, and the theory for dynamics of the relative rotation, sliding and screw like motion of nanotube walls has developed [22]. In this work, we show by molecular dynamics simulation that it is possible to construct a DWNT motor actuated by a DC voltage. The motor consists of a shuttle structure as shown in Figs. 4.11(a) and (b). Using classical molecular dynamics with empirical potentials, as detailed in Chap. 2, we show that the inner CNT can rotate.

Nanomotor Actuation

The shuttle nanotube based nanomotor can switch electrically using a low voltage. In our work we have investigated the rotation of inner nanotube after switching to the "ON" state. To explain the proprieties of nanomotor behavior, intershell electrostatic and van der Waals energies were investigated and inner CNT trajectories were studied using MDS. The charge distribution along CNTs is calculated as described in *phase* 3. Figure 4.13 shows the charge distribution on the carbon atoms of a roughly 5 nm long SWNT when the electric potential on the surface is 6.0 V. The electric charge value is in the range of 5×10^{-3} to 34×10^{-3e}. It can be seen that the charges on the atoms located at the central part of the tube do not vary significantly, whereas the charges on atoms at the two ends are much higher. The charge values reach up to 34×10^{-3e} for the open-ended nanotube. Figure 4.14(a) shows the electrostatic potential along a DWNT. It can also be seen that the tube ends are charged heavier. The electrostatic potential is much higher in the outer nanotube than in the inner one [25]. The potential difference is applied between two adjacent metallic contacts, and charges with opposing polarities are induced on two MWNT segments that are attached to these contacts as illustrated in Fig. 4.14(b). This charge distribution results in two types of electrostatic energies: an intra-segment repulsive energy (due to like charges) that extrudes the core from the outer housing (Fig. 4.14(b)), and an inter-segment attractive energy (due to charges with opposite polarities) that causes the core shells within each CNT segment to move towards each other

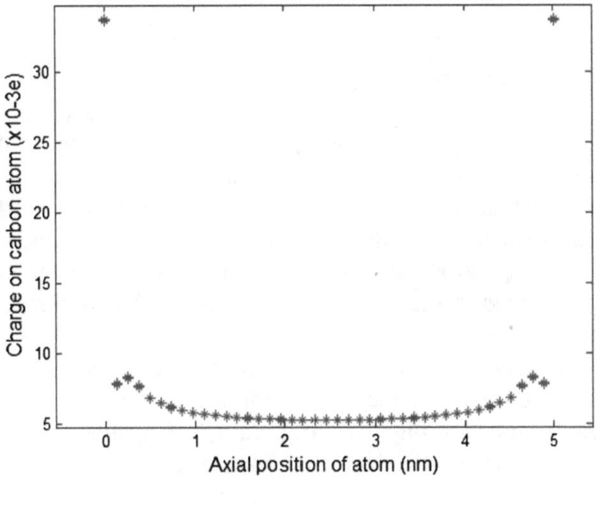

Fig. 4.13 Charge distribution along the axial direction for an open ended nanotube ($V = 6$ V)

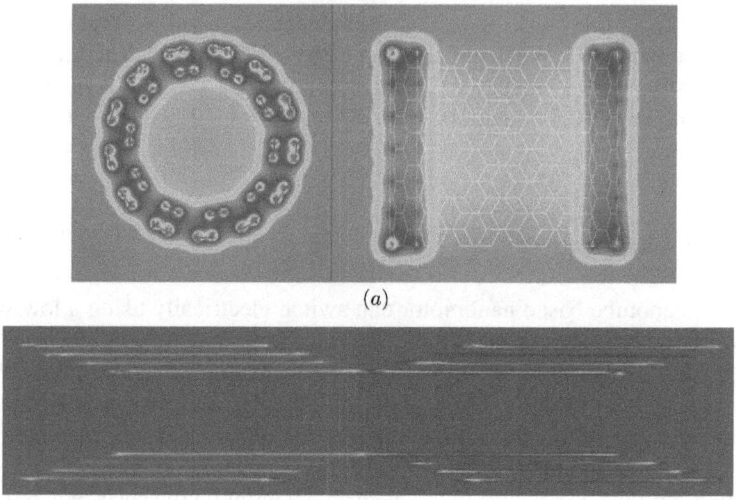

(a)

(b)

Fig. 4.14 (**a**) Electrostatic potential map along a DWNT structure. (**b**) Charge distribution during "turn-on" state of the negatively charged shells on the left and the positively charged shells on the right with cutaway view

(Fig. 4.14(b)). When the applied excitation bias exceeds a characteristic threshold value, the combined contributions of these two electrostatic force components will be higher than the van der Waals interactions within each of the two segments, and result in a closure of the gap. In this 'ON' state of the switch, a current flow is established between the metallic contacts. The attractive electrostatic energy between two oppositely charged inner nanotubes is shown in Fig. 4.15(a). Starting form 0.4 ns the two oppositely charged nanotubes move closer until the system turns on. Figure 4.15(b) depicts the repulsive electrostatic energy between the inner and the outer shells. The sliding time of the system is 0–0.4 ns.

Force Analysis During Rotation of Inner Shells

At the closed state of the system, a terminus atom in the inner shells nanotube is subjected to van der Waals forces and three components of electrostatic forces. Figure 4.16 shows a terminus carbon atom with applied forces. F_{eii} is an attractive electrostatic force which applied by the opposite inner nanotube. The two components F_{et} and F_{en} are the tangential and normal electrostatic repulsive forces which are applied by the outer nanotube. Van der Waals components was not shown in the figure.

In order to understand the origin of the inner nanotube rotation, we calculate the inter-layer interactions during rotation. We divide the nanotube into unit cells as illustrated in Fig. 4.18(a). Figure 4.18(a) shows the two types of energies acting on an inner unit cell as it approaches an outer one, these two energies are added

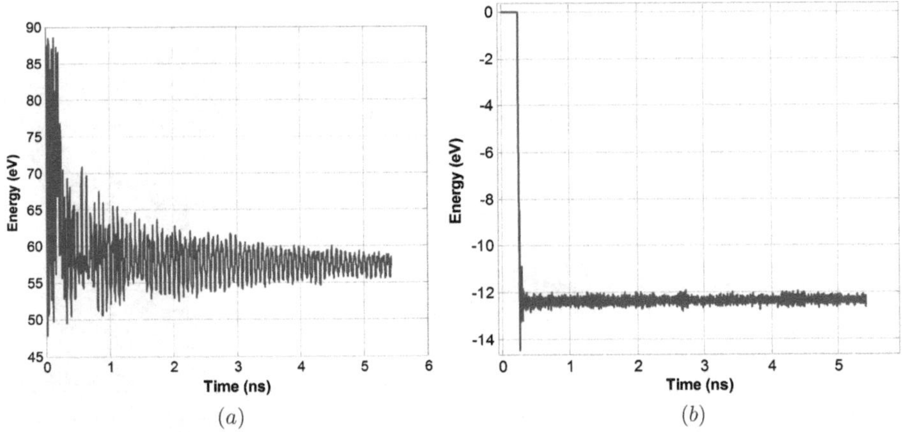

Fig. 4.15 (**a**) Attractive electrostatic energy between two oppositely charged inner nanotubes. Starting from 0.4 ns the two oppositely charged nanotubes move closer until the system turns on. (**b**) Repulsive electrostatic energy between the inner and the outer shells. The sliding time of the system is 0–0.4 ns

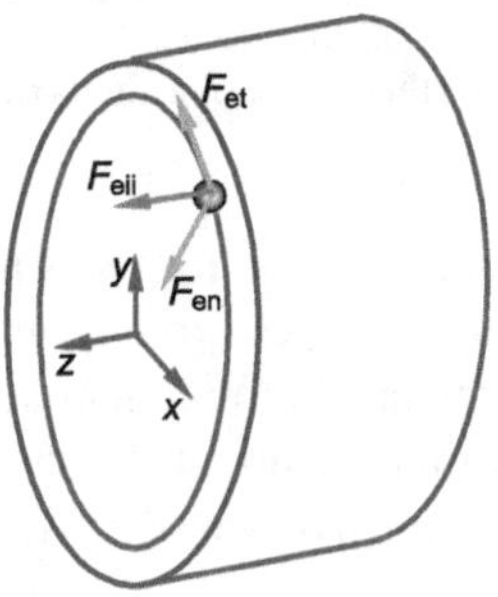

Fig. 4.16 Forces acting on a terminus inner nanotube atom. F_{eii} is the electrostatic attractive force from oppositely charged inner nanotubes, F_{et} and F_{en} are tangential and normal components applied by the outer nanotube

together to get the total non-bonded energy acting on the inner unit cell. The graph shows that from 2.780 ns to 2.818 ns the total attractive non-bonded energy (red curve) increases and from about 2.833 to 2.900 ns the total energy decreases when the two unit cells are getting close and separated. At about 2.828 ns the non-bonded energy decrease to −0.55 eV because of the repulsive van der Waals term when the atoms are too close. This shows clearly that the rotation of the inner nanotubes is mainly caused by the interlayer attractive non-bonded energy. We characterized the interaction energy between a unit cell in the inner and three successive unit cells in the outer nanotube during a rotation angle. Figure 4.18(b) illustrates the interaction between the three unit cells.

- At the time from 2.780 ns to 2.835 ns the total non-bonded energy between the inner segment (green colored) and the first outer segment (red) increases strongly. This means that the inner segment is strongly attracted by the first outer segment. At the time of about 2.828 ns the energy decrease to −0.55 eV which caused by the repulsive van de Waals energy when two segments become too close.

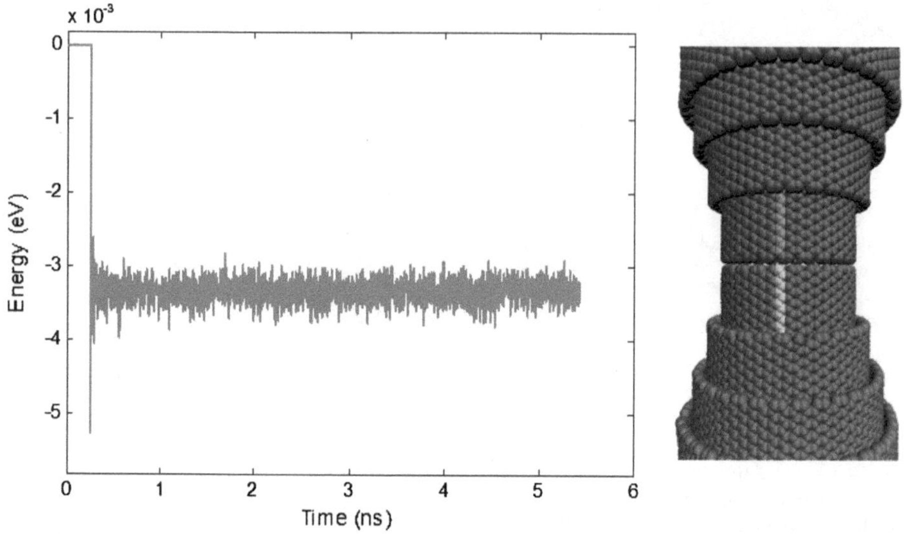

Fig. 4.17 (Color online) Electrostatic energy attraction between two opposite segments (*red* colored segment and *green* colored one) in inner shells. As shown by the curve, from 0 ns to 0.2 ns is the witching state, at 0.2 ns the head to head CNTs become closed, from 0.2 ns the energy is constant, which means that both inner shells rotate with the same velocity and in the same direction

- At the time from 2.820 ns to 2.848 ns the total non-bonded attractive energy between the inner segments (green) and the second outer segment (cyan) increases strongly. This means that the inner segment is strongly attracted by the first outer segment. At a time of about 2.835 ns the energy decrease to −0.55 eV which caused by the repulsive van der Waals energy when two segments become too closed.
- At the time from 2.830 ns to 2.850 ns the total non-bonded attractive energy between the inner segments (green) and the third outer segment (magenta) increases strongly. That means the inner segment is strongly attracted by the first outer segment. At the time of about 2.848 ns the energy decrease to −0.55 eV which caused by the repulsive van der Waals energy when two segments become too closed.

The calculations, using MD simulations coupled to electrostatic charge distribution calculations along the CNTs, show that electrostatic forces pulls head to head cores together (Fig. 4.15). The calculations show that the interlayer van der Waals forces at the "Turn on" state can generate a torque rotation as shown in Figs. 4.18(b) and 4.19(a). The Van der Waals forces dominate the electrostatic friction forces during rotation. The inner nanotube trajectory analysis shows that it is rotating at a circular path with constant velocity as illustrated in Fig. 4.18. The attractive electrostatic energy analysis between head to head CNTs shows that theses CNTs are rotating with the same velocity and in the same direction, as given by Fig. 4.17 the attractive electrostatic energy become stable after the structure turned on.

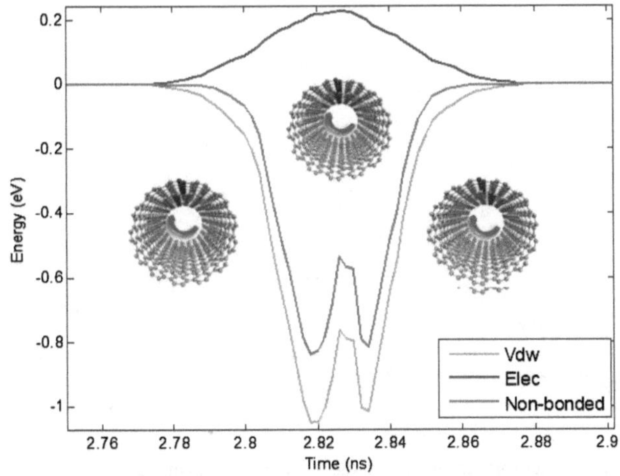

(a)

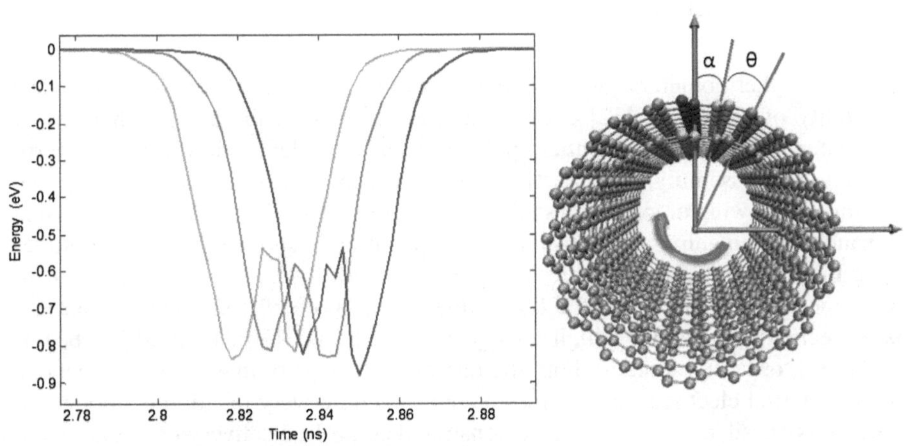

(b)

Fig. 4.18 (Color online) (**a**) Interaction energies between two segments in inner and outer nanotubes. *Blue curve*: electrostatic energy, *cyan curve*: repulsive van der Waals energy, *red curve*: total non-bonded energy. (**b**) Total non-bonded energy attraction between a segment with *green* color in the inner nanotube and three successive segments in the outer nanotube

4 Attogram Mass Transport and Vaporization Through Carbon Nanotube

Carbon Nanotube Boiler

Experiments were carried out in the Institute of Robotics and Intelligent Systems at the ETHZ in Switzerland by professor Bradley Nelson group [39, 40]. A variety of materials have been encapsulated by CNTs such as metals and their compounds [26], water [27], and fullerenes [28], and applications of devices as tem-

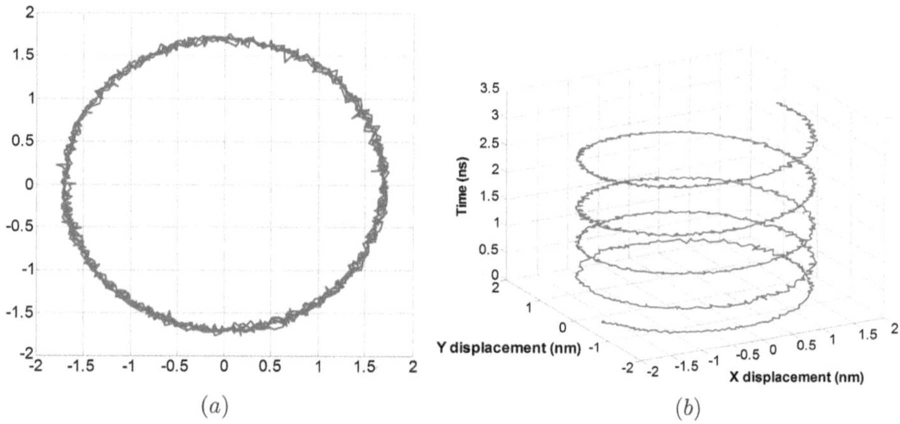

Fig. 4.19 Termini atom trajectory in an inner shell during rotation, (**a**) rotating circular path of inner CNT termini, (**b**) inner nanotube rotation versus time. The curve shows that the inner CNT rotates with constant velocity

plates [29], thermometers [30], and nano testtubes [31] have been described. The possibility of delivering [32] encapsulated material from the carbon shells is of great interest because of potential applications as atom-level sources for nanoprototyping, nanoassembly, and injection. An experimental investigation of controlled melting and flowing of single crystalline copper [33] from individual CNTs assisted by nanorobotic manipulation [34], and its application in spot welding of nanotubes using this copper were presented in a previous work [34]. A very low current induces the melting and drives the flow, which is far more efficient than irradiation-based techniques involving high energy electron beams [36], focused-ion beams (FIB) [37], or lasers. Because both the rate and direction of mass transport depend on the external electrical drive, precise control and delivery of minute amounts of material is possible, and we have demonstrated controlled delivery of attograms of mass from carbon shell barriers [35].

The experiments were performed in a transmission electron microscope (TEM, Philips CM30) equipped with a scanning tunneling microscope (STM) built in a TEM holder (Nanofactory Instruments AB, ST-1000) serving as a manipulator. The material consisting of a CNT bundle is attached to a 0.35 mm thick Au wire using silver paint, and the wire is held in the specimen holder (Fig. 4.20) [38]. Two kinds of probes are used. One type is an etched 10 μm thick tungsten wire with a tip radius of approximately 100 nm (Picoprobe, T-4–10–1 mm), and the other type is an atomic force microscope (AFM) cantilever (Mikromasch, CSC38/Ti-Pt). The probe can be positioned in a millimeter-scale workspace with sub-nanometer resolution with the STM unit actuated by a three-degree-of-freedom piezo-tube, making it possible to select a specific CNT and pick it up. Physical contact can be made between the probe and the tip of a nanotube or between two nanotubes. Applying a voltage between the probe and the sample holder establishes an electrical circuit through a CNT or a CNT-junction and injects thermal energy into the system via Joule heating. By increasing the applied voltage, the local temperature can be increased past the

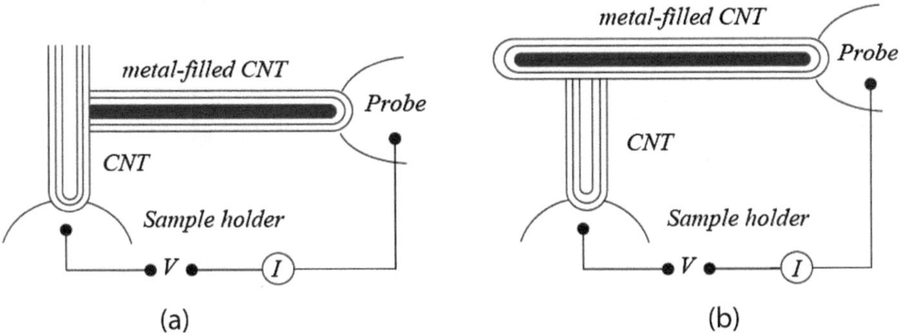

Fig. 4.20 Schematic setup of inter-nanotube mass transport (**a**) Cap-to-wall transport, (**b**) Wall–to-cap transport

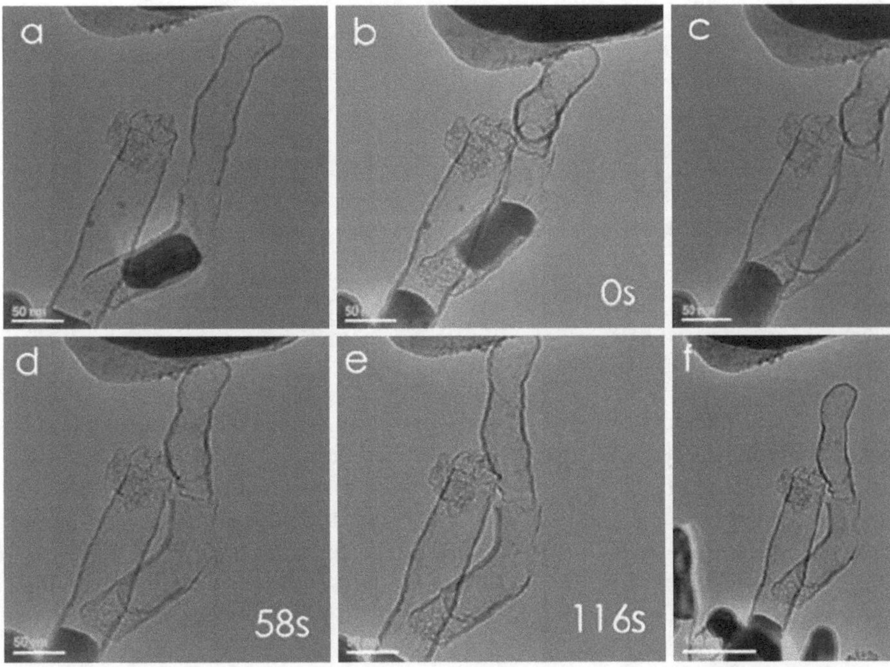

Fig. 4.21 Cap-to-wall inter-nanotube mass (Sn) transport. (**a**) A nanotube fluidic junction approaches a probe. (**b**)–(**e**) The process of inter-nanotube flowing. (**f**) The nanotube fluidic junction is detached from the probe. Scale bars: (**a**)–(**e**) 50 nm, (**f**) 100 nm

melting point of the material encapsulated in a tube. The process is recorded by TEM images, a multimeter, and a nA meter.

Figure 4.21 shows cap-to-wall inter-nanotube mass (Sn) transport. Figure 4.21(a) [38] is a TEM image of a nanotube fluidic junction, which includes two Sn-filled CNTs. The upper CNT has its opened cap attached on the walls of the lower tube.

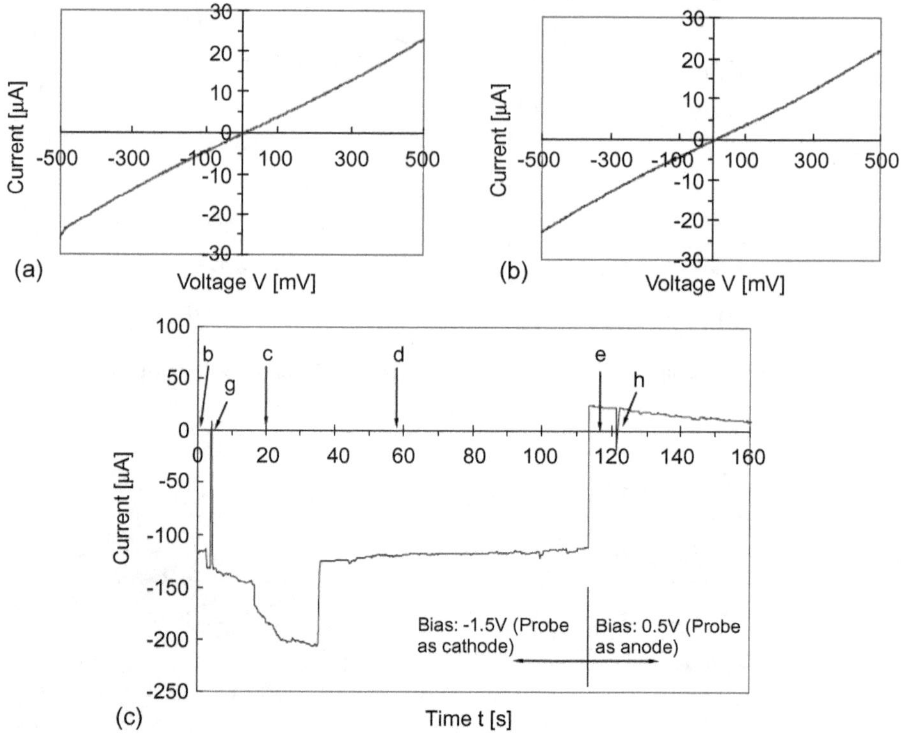

Fig. 4.22 (**a**) and (**b**) I–V curves recorded at $t = 4$ s and 122 s. (**c**) Current recorded with a multimeter (sampling rate: 4 Hz)

The junction is then attached to a probe. Parts (b) to (e) of Fig. 4.21 show the process of inter-nanotube mass flow. Figure 4.22(f) shows that the nanotube fluidic junction is detached from the probe. Figure 4.22(a), (b) and (c) [38, 39] are I–V curves recorded at $t = 4$ s and 122 s, respectively. It can be seen that no obvious changes occurred before and after flowing, suggesting that there is no significant altering of the carbon shells. Figure 4.22(c) [38, 40] is the current recorded with a multimeter (sampling rate: 4 Hz). It can be seen from the time scale the moments when the images (parts (b)–(e) of Fig. 4.21) and the I–V curves (parts (a) and (b) of Fig. 4.22) were taken. The amplitudes of the peaks in Fig. 4.22(c) for (a) and (b) are not as accurate as that shown in parts (g) and (h) of Fig. 4.21 [38] due to the low sampling rate. It can be seen from the sequence that as Sn entered the junction point ($t = 20$ s), the absolute value of the current increased, indicating Sn improved the contact between the two CNTs. Then a sharp drop appeared at 37.25 s, suggesting Sn left from the junction point. The current (absolute value) decreased slowly afterwards until the bias changed polarity at 116 s. This is due to the shortening of the Sn core, which exposes the carbon shells (and increases the resistance). After the polarity and value of the bias changed to 0.5 V, the current continues dropping implying a decrease of Sn on the contact between the lower CNT and the sample holder. According to the geometry and the density of Sn (7.31 g/cm^3), the original mass is approximately 662 ag (Fig. 4.21(b)), whereas the received mass by the lower tube is 377 ag.

It can be seen from the time scale the moment the images Fig. 4.21(b)–(e) [38] and the I–V curves Figs. 4.22(a) and (b) were taken. The amplitudes of the peaks in Figs. 4.22(c) for Figs. 4.22(a) and (b) are not as accurate as those shown in Figs. 4.22(a) and (b) due to the low sampling rate. It can be seen from the sequence that as Sn entered the junction point ($t = 20$ s), the absolute value of the current increased, showing Sn improved the contact between the two CNTs. Then a sharp drop appeared at 37.25 s, suggesting Sn left from the junction point. The current (absolute value) decreases slowly after until the bias changed polarity at 116 s. This is due to the shortening of the Sn core, which exposes the carbon shells (with larger resistance). After the polarity and value of the bias changed to 0.5 V, the current continues dropping implying a decrease of Sn on the contact between the lower CNT and the sample holder.

Molecular Dynamics Simulation

Using molecular dynamics simulation (MDS) we numerically investigated inter-tube mass (Cu) transport. Figure 4.23 [38] shows the initial atomic configuration of the simulated wall-to-cap transport system. The potential energy of the wall-to-cap transport system with a cluster of Cu inside was minimized at an internal pressure of 1 atm using the conjugate gradient method. To investigate copper diffusion, the system was simulated at temperatures between 700 K and 1800 K using molecular dynamics. The position of the copper ions as shown in the frames in Fig. 4.24 indicate that the copper has melted and diffused. The system temperature during diffusion is shown by the curve in Fig. 4.25(a) [38]. An analysis of the repulsive electrostatic energy between copper ions is given by Fig. 4.25(b) [38]. It can be seen that a peak temperature was reached at frame 200. Accordingly, as shown in Fig. 4.26 [38], the

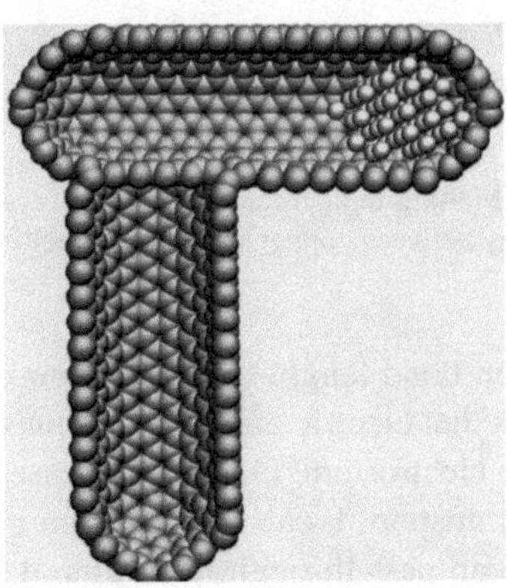

Fig. 4.23 (Color online) Initial configuration of the simulated system with a cut view. Copper cluster is *yellow*

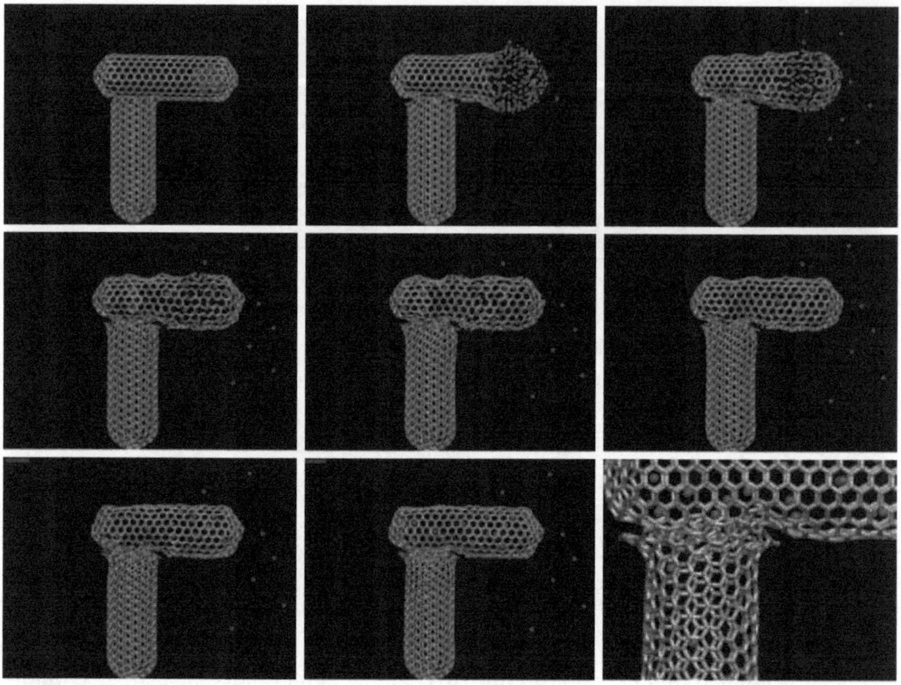

Fig. 4.24 Frames of melting and inter-tube diffusion of the Cu cluster using molecular dynamics simulation

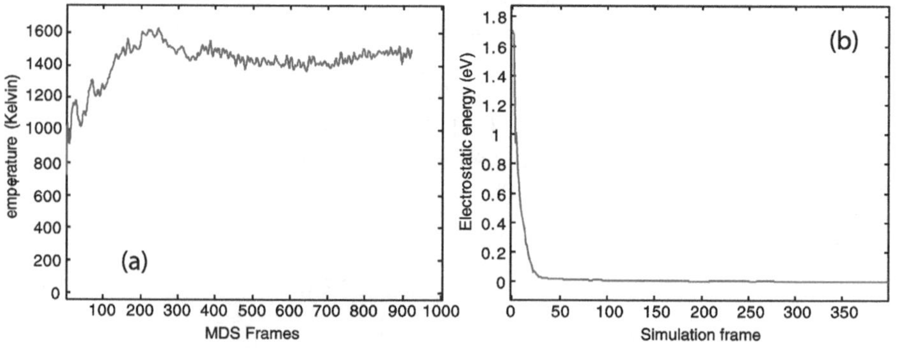

Fig. 4.25 (a) Melting temperature of copper crystal during simulation and (b) the electrostatic repulsive energy between copper ions during melting

carbon-carbon bond length obtained a maximum value at frame 200. The simulation indicates that electric energy is responsible for heating by joule effect, and that the repulsive electrostatic charges increase the distances between copper ions and induce their transport. It can be seen from Fig. 4.26 that with a large enough charge, copper ions can pass through the walls of CNTs without necessarily breaking the

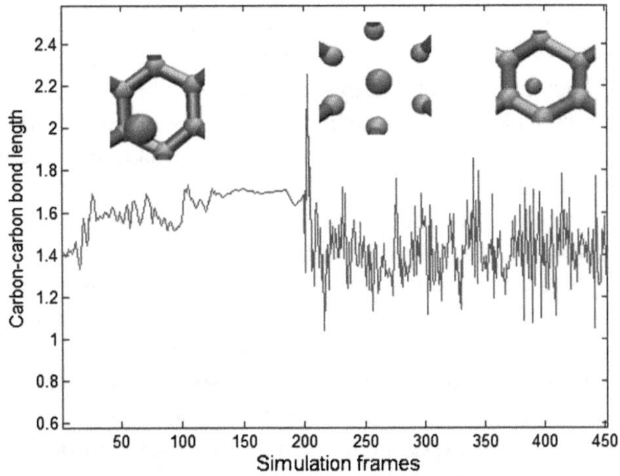

Fig. 4.26 Carbon-carbon bond length during copper diffusion. At frame 200 the copper ions pass through the hexagonal rings, which correspond to the maximum opening of the carbon rings crystal during simulation

bonds. Images clearly show that the hexagon carbon rings stretch during diffusion. Simulation also revealed that electrostatic forces guide the motion of the copper ions, causing the ions to collect in the original empty tube. It can be deduced that by increasing the number of shells, mass lost through the walls can be decreased.

In summary, Molecular Dynamic simulation of inter-tube copper transport indicates that electrostatic charge increases the distance between copper ions and then induces the transport of copper ions. Copper ions pass through the walls of CNTs without necessarily breaking the bonds, and the hexagonal carbon rings stretch during diffusion.

5 Conclusion

NEMS research involves design prototyping, fabrication, control, programming, and applications. Long time before the mass production of any nanodevice, prototyping is needed to ensure that the device will work as intended, and to optimize its parameters. Practical methodologies for setting up physically-based and consistent nanorobotic structure has been investigated through a co-prototyping methodology. It is an iterative optimization process coupling experiments performed on real multiwalled carbon nanotube-based nanodevices with computational methods of simulation. These experimental studies have been investigated in close collaboration with Prof. Bradley J. Nelson and Dr. Lixin Dong from the Institute of Robotics and Intelligent Systems at ETH Zurich in Switzerland. The co-prototyping work-flow is divided into 6 steps (1 to 6) with an interactive communication process between experiments and simulation data.

- The first step (step 1) in the work-flow is to define the first generation nanodevice for all further exploration. Each component in the structure is defined by a handful of parameters defining the size, position, orientation and materials properties.

Together these properties for all the materials make up the degrees of freedom of the system that can be optimized.

- The steps 2 and 3 provide to the designer an accurate physics-based model directly identified and parameterized by experimental measurements. The aim of the first two steps is to define, in a realistic and accurate way, the system being studied.
- Then, intermediate steps 4 and 5 initiate a co-optimization approach (coupling AIREBO force fields to experimental measurements) that can be used by itself to solve unconstrained problems or energy optimization. Model problems coupled to experiments demonstrate that the co-optimization method is computationally efficient and that it can be used to solve realistic problems (optimization of electrical potential w.r.t. gap distance, shell number and nanotube diameter).
- Finally, step 6 allowed the observation of certain phenomena which are not discernible by experiments such as the rotation of inners "head-to-head" nanotubes. It led to the design, simulation and characterization of rotational nanomotors. Furthermore, it allowed to explain an observable effect of fluid transportation inside nanotube junctions. Dynamic simulation of intertube copper transport indicates that charge increases the distance between copper ions and then induces the transport of copper ions.

In this modeling framework, the author demonstrates that among its' many benefits, the co-prototyping approach lowers the development cost, the development time involved in prototyping, allows for more iterations during experimentation, and gives developers the chance to get immediate experimental feedback on refinements to the design. The approach we also address some important issues to be focused on for the reliable validation and accurate calibration of the nanoscale models for accurate prototyping.

References

1. Cumings, J., Zettl, A.: Low-friction nanoscale linear bearing realized from multiwall carbon nanotubes. Science **289**, 602 (2000)
2. Kis, A., Jensen, K., Aloni, S., Mickelson, W., Zettl, A.: Interlayer forces and ultralow sliding friction in multiwalled carbon nanotubes. Phys. Rev. Lett. A **97**, 025501 (2006)
3. Stuart, D.J., Tuteinb, A.B., Harrisonc, J.A.: A reactive potential for hydrocarbons with intermolecular interactions. J. Chem. Phys. **112**(14), 6472–6486 (2000)
4. Cumings, J., Zettl, A.: Localization and nonlinear resistance in telescopically extended nanotubes. Phys. Rev. Lett. **93**, 086801 (2004)
5. Forro, L.: Beyond Gedanken experiments. Science **289**(5479), 560–561 (2000)
6. Dong, L.X., Nelson, B.J., Fukuda, T., Arai, F.: Towards nanotube linear servomotors. IEEE Trans. Autom. Sci. Eng. **3**, 228–235 (2006)
7. Yan, Q.M., Zhou, G., Hao, S.G., Wu, J., Duan, W.H.: Mechanism of nanoelectronic switch based on telescoping carbon nanotubes. Appl. Phys. Lett. **88**, 173107 (2006)
8. Maslov, L.: Concept of nonvolatile memory based on multiwall carbon nanotubes. Nanotechnology **17**, 2475 (2006)
9. Li, C., Chou, T.-W.: Theoretical studies on the charge-induced failure of single-walled carbon nanotubes. Carbon **45**, 922–930 (2007)

10. Subramanian, A., Dong, L.X., Tharian, J., Sennhauser, U., Nelson, B.J.: Batch fabrication of carbon nanotube bearings. Nanotechnology **18**, 075703 (2007)
11. Lozovik, Y.E., Minogin, A.V., Popov, A.M.: Phys. Lett. A **313**, 112 (2003)
12. Charlier, A., McRae, E., Heyd, R., Charlier, M.F., Moretti, D.: Carbon **37**, 1779 (2000)
13. Charlier, J.-C., Michenaud, J.P.: Phys. Rev. Lett. **70**, 1858 (1993)
14. Lozovik, Y.E., Popov, A.M.: Chem. Phys. Lett. **328**, 355 (2000)
15. Yoshida, M., Osawa, E.: Fuller. Sci. Technol. **1**, 54 (1993)
16. Lozovik, Y.E., Popov, A.M.: Phys. Solid State **44**, 186 (2002)
17. Lozovik, Y.E., Nikolaev, A.G., Popov, A.M.: Fuller. Nanotub. Carbon Nanostructures **14**, 227–231 (2006)
18. Dong, L.X., Subramanian, A., Nelson, B.J.: Carbon nanotubes for nanorobotics. Nano Today **2**, 12–21 (2007)
19. Subramanian, A., Dong, L.X., Tharian, J., Sennhauser, U., Nelson, B.J.: Batch fabrication of carbon nanotube bearings. Nanotechnology **18**, 075703 (2007)
20. Tu, Z.C., Hu, X.: Molecular motor constructed from a double-walled carbon nanotube driven by axially varying voltage. Phys. Rev. B **72**, 033404 (2005)
21. Belikov, A.V., Lozovik, Y.E., Nikolaev, A.G., Popov, A.M.: Double-wall nanotubes: classification and barriers to walls relative rotation, sliding and screwlike motion. Chem. Phys. Lett. **385**, 72–78 (2004)
22. Saito, R., Matsuo, R., Kimura, T., Dresselhaus, G., Dresselhaus, M.S.: Anomalous potential barrier of double-wall carbon nanotube. Chem. Phys. Lett. **348**, 187–193 (2001)
23. Schoen, P.A.E., Walther, J.H., Arcidiacono, S., Poulikakos, D., Koumoutsakos, P.: Nanoparticle traffic on helical tracks: thermophoretic mass transport through carbon nanotubes. Nano Lett. **6**, 1910–1917 (2006)
24. Tu, Z.C., Hu, X.: Molecular motor constructed from a double-walled carbon nanotube driven by axially varying voltage. Phys. Rev. B **72**, 033404 (2005)
25. Zólyomi, V., Koltai, J., Rusznyák, Á., Kürti, J., Gali, Á., Simon, F., Kuzmany, H., Szabados, Á., Surján, P.R.: Intershell interaction in double walled carbon nanotubes: charge transfer and orbital mixing. Phys. Rev. B **77**, 245403 (2008)
26. Ajayan, P.M., Iijima, S.: Nature **361**, 333–334 (1993)
27. Supple, S., Quirke, N.: Phys. Rev. Lett. **90**, 214501 (2003)
28. Smith, B.W., Monthioux, M., Luzzi, D.E.: Nature **396**, 323–324 (1998)
29. Ajayan, P.M., Stephan, O., Redlich, P., Colliex, C.: Nature **375**, 564–567 (1995)
30. Gao, Y.H., Bando, Y.: Nature **415**, 599–599 (2002)
31. Ugarte, D., Chatelain, A., de Heer, W.A.: Science **274**, 1897–1899 (1996)
32. Svensson, K., Olin, H., Olsson, E.: Phys. Rev. Lett. **93**, 145901 (2004)
33. Tao, X.Y., Zhang, X.B., Cheng, J.P., Luo, Z.Q., Zhou, S.M., Liu, F.: Diam. Relat. Mater. **15**, 1271–1275 (2006)
34. Dong, L.X., Arai, F., Fukuda, T.: IEEE-ASME Trans. Mechatron. **9**, 350–357 (2004)
35. Dong, L.X., Tao, X.Y., Zhang, L., Zhang, X.B., Nelson, B.J.: Nano Lett. **7**, 58–63 (2007)
36. Dong, L.X., Arai, F., Fukuda, T.: Appl. Phys. Lett. **81**, 1919–1921 (2002)
37. Matsui, S., Kaito, T., Fujita, J., Komuro, M., Kanda, K., Haruyama, Y.J.: Vac. Sci. Technol. B **18**, 3181–3184 (2000)
38. Dong, L.X., Tao, X.Y., Hamdi, M., Zhang, L., Zhang, X.B., Ferreira, A., Nelson, B.J.: Nanotube fluidic junctions: inter-nanotube attogram mass transport through walls (2008, submitted)
39. Dong, L.X., Tao, X.Y., Zhang, L., Zhang, X.B., Nelson, B.J.: Metal-filled carbon nanotubes for nanomechatronics. In: Proc. of the 2008 IEEE/ASME Int'l Conf. on Advanced Intelligent Mechatronics (AIM2008), Xi'an, China, pp. 933–937 (2008)
40. Dong, L.X., Tao, X.Y., Zhang, L., Zhang, X.B., Nelson, B.J.: Nanorobotic spot welding: controlled metal deposition with attogram precision from copper-filled carbon nanotubes. Nano Lett. **7**, 58–63 (2007)

Chapter 5
Conclusion and Future Prospects

1 Conclusion

Nanorobotics is an emerging area of scientific and technological opportunity. It is a new and rapidly growing interdisciplinary field addressing the assembly, construction and utilization of molecular devices based on nanoscale principles and/or dimensions. The size-related challenge is the ability to measure, manipulate, and assemble matter with features on the scale of 1–100 nm. A nanorobot is essentially a controllable machine at the nanometer or molecular scale that is composed of nano-scale components and algorithmically responds to input forces and information. Complex nanodevices can be devised and designed through biomimetics. Bio-nanorobotics, namely biomolecular robots, represents a specific class of nanorobots where proteins and DNA could act as motors, mechanical joints, transmission elements, or sensors. If all of these different components were assembled together they can form bio-nanorobots with multi-degree-of-freedom, able to apply forces and manipulate objects in the nanoscale world. These bio components seem to be a very logical choice for designing nanorobots. In this work, we investigated the design, assembly, simulation, and prototyping of biological and artificial molecular structures with the goal of implementing their internal nanoscale movements within nanorobotic systems in an optimized manner.

By reviewing the state-of-the-art in the domain, Chap. 1 has shown that although many of the described nanorobotic technologies, molecular and atomistic modeling algorithms, virtual reality based simulation environments have been developed into more or less mature products for robots acting in the meso- and micro-world, the nano-size of the robots poses extreme challenges and requires a complete rethinking of nanorobotics design and control. Designing nanorobotic systems deal with a vast variety of sciences, from quantum molecular dynamics, to kinematics analysis with emphasis on advanced motion control. To facilitate user input in nanorobotic systems it is essential to develop virtual reality based design techniques using visual and haptic perception coupled to molecular modeling algorithms. Using simulated nano environments in VR, the operator can design and characterize through physical simulation the behavior of nano- and bio-nanorobots. Adding haptic interaction,

M. Hamdi, A. Ferreira, *Design, Modeling and Characterization of Bio-Nanorobotic Systems*,
DOI 10.1007/978-90-481-3180-8_5,

the operator can explore and prevent the problems of nano and bionano-robots in their native environment. Based on these developments, we discussed several design rules, molecular and quantum molecular methodologies, empirical force fields and interactive VR techniques which could be applied in a new design philosophy and architecture for nanorobotic systems.

The roadmap for the development of future bio-nano and nano components involved the development of nanorobotic design and prototyping methodologies. In Chap. 2, we proposed two CAD-design approaches. The first one involved multiscale modeling tools (quantum mechanics, molecular dynamics, continuum mechanics) coupled to virtual reality advanced techniques. In order to design and evaluate the characteristics of molecular robots, we proposed interactive nanophysics-based simulation which permits manipulation of molecules, proteins and engineered materials in molecular dynamics simulations with real-time force feedback and graphical display. Communication is achieved through an efficient socket connection between the visualization package and a molecular dynamics and quantum mechanics tools running on single or multiple machines. This approach allows conceptual representations of modular organization of a bio-nanorobots. It consists in the development of new design methodologies, assembly of functionally stable bio-nano components (design connectors for binding the bio-nano components) into complex assemblies through energy minimization algorithms. The second approach uses a novel co-prototyping methodology. The optimization of engineered nanorobotic device is coupled to experimental measurements and force field modeling algorithms. In some cases, the design parameters are not accessible by experimental measurements which leads the designer to not be able to characterize deeply the nanodevice operating principles leading to optimal performances. The flow optimization process is evaluated with respect to accessibility of experimental measurements, parametric model optimization, design constraints, and device operation limitations. Experimental and simulation data are the key points of the co-prototyping method. With these both design methodologies in mind, we have become interested in developing two kinds of nanodevices, i.e., (i) interactive multiscale design of protein-based nanorobotic components using the mechanical force and movement generated by molecular proteins, and (ii) co-prototyping of multi-walled carbon nanotube based nanodevices using electrostatic intershell interaction.

As a substantial step toward the long-term objective of creating protein-based nanomechanical devices, we have demonstrated different "proof-of-concept" nanomechanical devices driven by the protein and/or DNA molecules. The mechanical properties of these bio-nanodevices can be controlled by using rational design based on protein engineering principles. First, the development of bio-nano components from biological systems is the first step towards the design and development of advanced bio-nanorobots. Simple protein-like elastic joints elements for structural links have be carefully studied and simulated through steered molecular dynamics to understand their functional limits. Through this *library of bio-nano components*, complex assemblies have been prototyped using interactive design, multiscale modeling, and molecular control. This work has been realized in close collaboration with the Biomedical Mechatronics Laboratory, Dept. of Mechanical and Industrial Engineering, University of Northeastern (USA). We have demonstrated that realistic design and simulations can be carried out by integrating the physics at several scales (atomistic, nanometer, molecular and mesoscopic) for various times scales of simulation (from few nanoseconds to several milliseconds). As example, dynamic characteristics of molecular sarcomere mechanism composed of proteins-based passive kinematic chains and actuated by myosin protein-based nanoactuator has been successfully simulated. We demonstrated also that DNA material is well suited to the development of biomotors for the control of nanorobotic tasks such as molecular linear biomotors and nanotweezer working in biological mediums. We demonstrated that these characteristics are strongly dependent on medium properties: temperature, protonation and electron density control parameters. Further development in the prototyping strategy will be to integrate standard control softwares (namely, MATLAB Simulink) in the multiscale simulation platform in order to evaluate the performances of advanced controllers (robust, adaptive and stochastic) for closed-loop nanorobot control prototyping.

Finally, co-prototyping design methodology has been successfully validated through the optimization of multiwalled carbon nanotube-based nanodevices in close cooperation with the Institute of Robotics and Intelligent Systems, ETH Zurich (Switzerland). To demonstrate the effectiveness of the approach, we used jointly atomistic modeling tools and experimental measurements to support the designer in the structure design, performance characterization, and energy optimization. The proposed simulation methods used *ab-initio* quantum mechanical methods and classical molecular dynamics with AIREBO potential. Design, simulation and optimization of high-density electromechanical switches based on bi-directional linear bearing has been has been concretely applied. Furthermore, we demonstrated that co-prototyping methodology is a valuable tool to demonstrate new atomistic physical effects such as nanotube fluidic junctions for the internanotube attogram mass transport through walls. A molecular dynamics simulation of the physical effect has been presented that provided further insight into the transport mechanism. Simulation also revealed that electrostatic forces guide the motion of the copper ions, causing the ions to collect in the original empty tube. It can be deduced that by increasing the number of shells, mass lost through the walls can be decreased. Hence, an optimization of the number of shells will be needed when designing a fluidic system for a specific application. We conclude that cooperative work between experiments and simulation paves the way for high-performance nanorobotic component designs.

2 Future Prospects

While a more careful evaluation of the proposed methodologies are ongoing and will be continued in the future, the purpose of this study is to demonstrate the capabilities of the design and simulation tool set. The author believes that this work is a huge proposition and that only an example among many potential possibilities is described to explain the concept. Further work with continuous upgrades will be required as the field of research progresses, notably to perform a faster development and control of nanodevices based on different nanomaterial structures: graphite mono or multilayer, electronic materials, proteins or lipids. A necessary work will be to characterize experimentally some of the proposed designs. The future prospects calls for collaborative efforts between physicists, chemists, biologists and nanotechnologists.

- First, a collaboration is underway with the Centre de Biology Moléculaire–CNRS (University of Orléans) in order to experiment DNA-based active and passive nanorobotic elements. The modulation of the strength of the DNA based actuator will be considered. In this regard, force control using variable DNA stiffness can be realistically envisaged since previous experiments on DNA overstretching have shown that effects of solution conditions plays an important role in force-extension curves. Variations of the medium environment have been controlled through three parameters: (i) the temperature, (ii) the level of acidic pH and (iii) the electron density of the molecule.
- Second, novel nanocapsules-based CNT designs with active drug delivery systems can be prototyped and optimized for targeted drug delivery. In the framework of the FP7-ICT-2007-2 NANOMA project (Nano-Actuators and Nano-Sensors for Medical Applications) coordinated by the Institut PRISME, University of Orléans, further upgrades of the multiscale simulation platform are forseen. As example, the designer will be able to design nanocapsule structures coated with specific recognition ligands (peptides, glycoproteins) that bounds to the cell surface in order to perform a site selective delivery, to conceptualize some generic scenarios of nanocapsules and biomolecular interactions for efficient drug delivery, and to control its intravascular motion through simulated high magnetic gradients.

In summary, the future of computational nanotechnology is bright, and strongly correlated with the growth and developments in bio-nanotechnology. A brief review of the current status, on-going applications and future prospects covered in this thesis shows that large scale modeling and simulation will not play a critical role in the gradual advances, but also may provide some revolutionary concepts for the novel technologies in a wide variety of nanotechnology related areas.

FSC
www.fsc.org
MIX
Papier | Fördert
gute Waldnutzung
FSC® C083411

Kolibri 360 GmbH
Ferdinand-Jühlke-Straße 7
99095 Erfurt, Deutschland
produktsicherheit@kolibri360.de